The Plant Protection Discipline

THE PLANT PROTECTION DISCIPLINE

Problems and Possible Developmental Strategies

by

Webster H. Sill, Jr.

ALLANHELD, OSMUN & CO.
MONTCLAIR

A Halsted Press Book
JOHN WILEY & SONS NEW YORK CHICHESTER BRISBANE TORONTO

ALLANHELD, OSMUN AND CO. PUBLISHERS, INC.
19 Brunswick Road, Montclair, N.J. 07042

Published in the United States of America in 1978
by Allanheld, Osmun & Co.
Distribution: Halsted Press,
a division of John Wiley & Sons., Publishers,
605 Third Avenue, New York, New York 10016

LIBRARY OF CONGRESS CATALOGING IN PUBLICATION DATA

Sill, Webster H., Jr. 1916 -
The plant protection discipline.

Bibliography: p.
1. Plants, Protection of. I. Title
SB950.S55 632 78-59171
ISBN 0-470-26443-8

Printed in the United States of America

Dedicated to Charlyn Sill
a patient and loving wife who
sat at home alone numerous nights
while this book was being written

Acknowledgments

Much of this work was done while the author was a Senior Fellow at the Food Institute of the East-West Center in Honolulu, Hawaii, on leave of absence from his regular posts of Chairman, Biology Department, University of South Dakota at Vermillion, and Director, Center for Environmental Studies, State of South Dakota.

The author particularly appreciates the permission to go on leave of absence granted by President Richard Bowen and Dean Donald Habbe, Arts and Sciences, of the University of South Dakota. Hearty thanks are also due Dr. Donald Dunlap who served as Acting Chairman in the author's absence, and to a patient and considerate wife who often put up with an absentee husband.

At the Food Institute, several have been most helpful in this work. Dr. Philip Motooka, Research Associate in Pest Management, has given many suggestions and much valuable help of many kinds. Dr. Nicolaas Luykx, Director of the Food Institute, has been a source of much appreciated inspiration and advice. Mrs. Fannie Lee-Kai and her faithful secretarial corps have been most efficient in typing and editing. Mary H. Larson was most efficient and helpful in typing and proofreading and deserves special thanks. Mrs. Rita Hong and her helpers have been most helpful in obtaining the necessary books and periodicals for study and perusal. Finally, the author particularly appreciates the granting of the position of Senior Fellow which was granted him through the approval of Dr. Everett Kleinjans, President of the East-West Center, and his Board of Governors.

This work does not reflect the official position of the Food Institute of the East-West Center. The author bears sole responsibility for the contents, both the organization and presentation of the ideas developed in the text. The author will admit to a plant pathological bias, but he has attempted to be completely fair with all of the subdisciplines of Plant Protection. Some will feel that fairness has been achieved; others may not. Any errors or misinterpretations must be attributed to the author alone. It is earnestly hoped that this work will prove useful to those interested in the future organization of cooperative efforts in teaching, research and extension in the evolving Plant Protection discipline.

Webster H. Sill, Jr.
Chairman, Biology Department
University of South Dakota
Vermillion, South Dakota
April 28, 1978

Table of Contents

The Plant Protection Discipline

Chapter 1

Introduction

The clash of doctrines is not a disaster; it is an opportunity.
WHITEHEAD

For over 15 years, the author has been a quiet observer of developments in crop protection and pest management in the United States. The leaders in initiating developments and various definitions of pest management have been largely entomologists. In opposition to this group have been mostly the plant pathologists who have been fostering the concepts of crop protection and plant health. Both groups have been pushed into action by environmentalists attempting to curtail the use of pesticides. Both entomologists and plant pathologists have now become interested in developing control procedures for pests, diseases, and maladies of plants which reduce or eliminate our dependency on pesticides.

This work is an effort to document that conflict: to show not only what has happened, but also what the present worldwide trends are in the broad discipline of plant protection (crop protection and plant pest management). The philosophies and points of view of the various positions will be explored.

This is not a scientific treatise. Rather it is a compendium and summary of the thoughts of the leading proponents or opponents of this new discipline. Consequently, there are many quotations, far more than most scientists, including the author, might prefer. However, it was felt that each advocate had a right to argue his own position. Much of this controversy is not to be found in the literature. Hence, many of the quotations contained in this book are excerpts from personal correspondence with the author. A few are anonymous where particularly sensitive matters are discussed, or where this was requested.

This work is designed primarily for government officials, administrators, and leaders in the various subdisciplines. Hopefully, it will bring the whole discipline of plant protection into a more realistic worldwide perspective. Certain important points are emphasized and re-emphasized even at the risk of being occasionally repetitive. In this discussion the stakes are too high for us to indulge in stylistic niceties. What we need now and need urgently is the development of a cooperative, practical, worldwide program of plant protection.

The environmental movement has pushed entomologists, in particular, in the direction of integrated pest management. Plant pathologists and nematologists have either been utilizing these integrated techniques in the past or have gradually been moving in recent years toward the development of more integrated or combined control procedures. This is happening because of several mounting pressures: the increased cost of pesticides, the development of pests resistant to chemicals, the mass destruction of useful insects and mites by some pesticides, the danger of some pesticides to man and domestic and other warm-blooded animals, and the continued use of certain wide-spectrum pesticides that appear to accumulate to dangerous levels in the environment (Wood *et al.* 1969). From a worldwide standpoint the overriding consideration pushing crop protection and pest management programs toward cooperative efforts and greater efficiency is the human population explosion and the food requirements for the future that are posed by that problem.

Palm (1972) summarizes this frightening population problem and its relationship to plant protection:

> The projection for doubling the human population by the beginning of the next century is so staggering as to be scarcely comprehended by the average person. Because of this, it is essential, then, that scientists interested in insect pest management initiate a set of national goals and work to attain them. While the entomologists are joining forces, we must realize that other scientists in the fields related to the control of plant diseases, vertebrate pests, nematode pests, weed pests, and others are facing equal or greater challenges than the entomologists. Our concern (in entomology) is directed toward insect pest management, yet we need to be alert to fit our programs into the broad demands for pest control in relation to environmental quality.

What is not so widely known is that at least half of the world's critically short food supply is destroyed each year, by plant diseases, natural disasters, insects, rodents, birds, nematodes, and other pests that attack foods not only in the field, but also during shipment and storage. The fastest and least costly way of greatly increasing the available food supply is to control diseases and pests in storage and shipment. If this were done, there would be an immediate 25 percent increase in edible food grains without any alteration in basic agricultural productivity. If, in addition, the pathogens and pests that attack the principal cereal grains were controlled more adequately, there would be approximately an additional 200 million tons of grain available each year, enough to feed about one billion people. Even in the United States, where great efforts are being made to control pests, about one quarter of our food is lost to insects, diseases, weeds, and other pests (Brady 1974).

In poorer countries, food losses are even greater. In India, for example, 70 percent of all food in storage is destroyed by diseases and pests. A half-dozen rats eat as much grain as a man. Rats also store large quantities of their own food reserves. Rodent burrows are often found with as much as ten pounds of grain in reserve storage (Brady 1974).

The nutritional value of grain also deteriorates sharply because of attacks by insects and infections associated with microorganisms, particularly fungi. Often pests preferentially consume the protein-containing portions of grain, thereby reducing its nutritional quality (Brady 1974).

Food shortages, owing to destruction by diseases and pests, often provoke extreme political and social consequences that have altered the course of history. The classical example, of course, is the potato blight in Ireland and northern Europe which caused the mass migration of Irish and many Germans to the United States in the middle 1800's. The grape mildew of the middle 1800's, particularly in France, is another classical example of a crop disease that caused extreme political and social problems associated with the near collapse of the wine industry.

In the developing world of today, crop losses owing to plant diseases, pests, drought and natural disasters are actually so staggering that political regimes may collapse because of the damages caused. A solution to the storage problems associated with small grains alone could probably help to stabilize the political and social climate in many countries of the developing world.

Although cooperation between various plant protection disciplines is badly needed, such cooperation is certainly not characteristic of the developed world at present and often is absent in the developing world. The various disciplines associated with plant protection developed separately as natural disciplines. These disciplines evolved largely over the past 100 to 150 years and only recently have produced excessive specialization. However, the many specialities must meet at the field level, as field extension personnel have always known. Russell (1955) warned of the danger of the fragmentation of the disciplines:

> The modern danger is that specialists may lose touch with each other and the subject may end up as it began; as a mass of facts of

> greater or less moment, out of which a clear and lucid system is yet to be framed.

Russell felt that the dangers could be averted,

> only by arranging regular meetings for the reading and discussion of papers and carefully planned conferences where subjects of special importance can be examined on broad lines.

There certainly has been no organized effort to avoid cooperative research or to avoid the broader problems which appear at the field level. It has simply been more convenient, and often easier, to work on smaller specialized problems that could be analyzed by known research techniques.

The computer has made it possible to consider some of the general, interconnected, and broad field problems. Now it is possible to track a complex network of variables via the computer. Heretofore, such problems had to be solved empirically, if at all, by applied or adaptive research consisting largely of statistical analysis of carefully designed field plots.

As every field worker can testify, recommendations by agronomists may often conflict with recommendations by plant pathologists or entomologists. Such conflicts can often be embarrassing, and are seldom aired in public. It is interesting to note that very few of these disagreements, which occur rather frequently at the field level, have been recorded in the literature, even though every worker who has functioned for long periods at the field level can give important examples of cases of conflicting recommendations.

Actually, the various plant protection disciplines are completely intertwined and interdependent with the agronomic and economic crop requirements at the field level. A farmer grows a crop and he is primarily interested in improvement, production, and management of that crop; that is, in the production of a healthy, high-quality, high-yielding crop. Unfortunately, his concern for plant protection usually does not begin until there is a problem so large that it can no longer be ignored from an economic standpoint. This is where the specialists in plant protection enter into the grower's thinking.

The world of the farmer in the field is very confusing and complex. Symptoms of plant diseases may often be confused with insect damage or nutritional deficiencies. There are climatic and weather problems which can be confused with pest problems. One seldom goes into the field and sees a simple clear-cut case of just one disease or one pest present in a crop. Usually there is a combination of diseases, pests, maladies and disasters, and the field worker is forced to analyze the symptoms and/or syndromes as best he can from the data available.

Problems of agriculture at the field level make all workers humble and the wisest quickly admit their basic ignorance. For this reason it is quite common for extension specialists representing the different crop disciplines to travel together to examine fields. Such teams of specialists accompanied by the grower can often develop some of the most creative approaches not only toward diagnosis but also toward needed research. Possibly the best beginning for a cooperative team research program in crop protection and pest management in a given crop is for individuals representing the different disciplines to travel together and look at field problems together with the grower in this most natural of situations. The interchange and flow of ideas in such a group is typically most useful and often highly creative in terms of its potential for new ideas and new approaches to understanding and possible control.

This acute need for an interdisciplinary approach, particularly at the field level, is summarized by Good (1974a):

> The need for interdisciplinary approaches to pest management in research, extension, and regulatory programs is fundamental to agriculture. Pests do not exist and act independently, nor can effective control programs be developed without considering the complex interrelations of crops, soil environment, and pests. We must recognize that effective pest management must be founded on sound principles and based on·adequate research. Effective pest management must recognize that pest problems have been created (largely) by the act of farming itself, and in recent years

> magnified by intensive agriculture and sometimes acute mismanagement. Most pest problems, whether they be weed, insect, nematode, disease, or a complex, can be reduced or eliminated as economic pests by properly managing farm units. Others will require community or even wide area suppression, or eradication programs. Regardless of the professional discipline, the basic approaches to pest management practices are the same, such as land preparation, precise planting and harvesting dates, post harvest crop destruction, water and fertilizer management, and pest free seed and plant propagating materials.

As Good has indicated, many of man's activities have favored the development of pests and plant diseases. Yarwood (1970) estimated that between 1926 and 1960 the number of recorded diseases of our principal crops increased about threefold. He gave the following reasons:

(1) Introduction of new plants, and the commercial movement of plants;
(2) Vegetative propagation of plants;
(3) Development of large monoculture systems of agriculture extending over vast acreages;
(4) Tillage techniques;
(5) Certain harvesting techniques;
(6) Certain storage techniques;
(7) Ideal fertilization, which tends to create more lush, succulent plants which are often more attractive to pests and more susceptible to diseases;
(8) Irrigation techniques and the movement of irrigation water;
(9) Use of herbicides;
(10) Plant breeding techniques that tend to narrow the gene base for resistance to a certain disease or pest;
(11) Location of plants on particular sites or in areas away from the best regions of adaptation;
(12) Disease-producing or damage-producing chemicals.

It would seem that if the above were actually true, all of these techniques listed should be eliminated or altered. What is not mentioned here is that despite the fact that these tend to stimulate the development of crop pests and diseases, it is still true that all of the above techniques stimulate much more the increases in plant production and yield in modern agriculture. The gains in productivity, typically, far outweigh the losses from pests and diseases. For these reasons alone, recommendations concerning crop improvement, production, and management, typically, will prevail over those which primarily concern plant diseases and pest control, especially where there is a conflict.

There are not only many interdisciplinary technological problems associated with plant protection but there are also increasingly important political and sociological problems which have often been ignored in the past. R. Smith (1969) emphasizes the social, and often political, ramifications of control efforts in crop protection and pest management:

> The agro-ecosystem is many things beyond the relationships among the crop plants and their conditioning environments. It also embraces the total associated agricultural, industrial, recreational, and social spheres. Hence, insect control must accommodate itself to the constraints of society. The habits, customs and traditions of the culture must be accommodated, religious beliefs, the structure of land tenure, marketing systems, and educational institutions can all help or hinder a technological measure or modify the magnitude of a pest problem. An action resulting in a new or altered technology may have significant social or political consequences. The economic and entomological aspects must be considered together through cooperation between entomologists and economists (this is a new role for both). The "extra-limital" aspects go well beyond the traditional boundaries of economics. They very much involve the social and political spheres.

Cotton production in Central America illustrates some of these possible political ramifications (R. Smith 1969):

> Insect pest control there is just one or two growing seasons away from a calamitous situation such as affected the Canate Valley of Peru in 1956. The bollworms, *Heliothis zea* and *H. virescens* have now replaced the boll weevil as the most important pest and show increasing evidence of resistance to the available pesticides. An array of formerly minor pests has been raised to major status. The dosages and numbers of applications of pesticides have been increasing steadily with poor control results. In

> several areas the average number of applications is over 30 per year, and some individuals have made over 50 treatments to a field in a single season. The possibilities of a cotton failure are very great in the next year or two if current practices are not modified. The possible social and political implications are many, especially when one considers that over 30 percent of the export dollars for countries like Guatemala and Nicaragua come from the sale of cotton fiber. It is no exaggeration to say that pest control advice which leads to an economic calamity may topple a government. The further complications are many and foreboding.

Much more attention needs to be given in the future to such political and social problems stemming from attempts to control plant pests and diseases. Until recently very little attention was given to the possibility of developing cheaper and more efficient cooperative strategies of crop protection and pest management (E. Smith 1972, and R. Smith 1972). As Robins (1972) points out, "The integration of all control technologies in a system of pest management" is what is actually needed. We need to develop a system of pest and disease management and prevention which maximizes the application of the total control components, including all that we know about control of pests and diseases of a given crop. Recently, there has been a concentration upon the development of insect pest management and integrated pest management and control systems. What we now need to do, according to Robins, is broaden the base of this concept to deal with the complete array of pests and diseases, including weeds, nematodes, plant diseases, rodents and all the rest.

In reality, we have a long, long way to go before many coordinated, cooperative, interdisciplinary control programs become a reality. This point is emphasized by Good (1974c):

> American farmers have yet to achieve comprehensive crop production and protection systems that are based on sound pest management and ecological principles and designed to control a broad spectrum of pests. Too often recommendations for the control of a specific pest, regardless of the method employed, are made without consideration of the management of other pests of local or area wide importance. Plant pathologists, entomologists, nematologists, and weed scientists have independently

> developed many pest management practices that reduce pest damage on specific crops but not for entire farming systems. The technology is now available to begin providing American farmers with better decision-making capability in regard to developing integrated systems to achieve optimum efficiency in production.

And even earlier, Good (1973b) urged the following:

> The objectives of the pest management programs in the United States are to establish multiple and alternate choice systems of pest control that are effective, economical, and environmentally sound. The ultimate goals of these projects are to promote effective use of combinations of cultural, biological, and chemical methods.

E. Smith (1972) also emphasized the importance of developing combined or coordinated methods of control.

> The prevailing concept at this time in pest control centers on population dynamics. From an understanding of the factors regulating pest populations we seek to invoke a number of practices that in total reduce the population to tolerable agricultural levels. This de-emphasizes dependence upon a single method and encourages utilizing more than one method to control a pest or disease where known.

Official guidance in the form of a policy statement by the Secretary of Agriculture came in 1973 and suggested a new movement in the direction of integrated control (Good 1973a):

> Non-chemical methods of pest control, biological or cultural, will be used and recommended whenever such methods are economically feasible and effective for control or elimination of pests. When non-chemical methods are not tenable, integrated control systems utilizing both chemical and non-chemical techniques will be used and recommended in the interest of maximum effectiveness and safety.

Thurston (1974) also emphasized the need for an integrated control approach as it relates to plant diseases.

> Plant disease control cannot focus only on the pathogen or the damage it causes. It has to be integrated with the entire package of management practices in agriculture, especially those related

to insect and weed control. The object of these practices is the production of maximum yield consistent with sound ecological principles which maintain or result in a wholesome environment. Plant disease control is an integral part of crop protection or pest management.

Leaders from all of the plant protection disciplines seem to agree that a concerted effort is needed in the direction of interdisciplinary, coordinated, cooperative, combined, and integrated control programs and the supporting research that is necessary to achieve these. Certainly, the problems are most complex and can only be solved by concerted effort by all of the disciplines working in concert. What the farmer needs and wants is something very inexpensive, almost automatic, which can be handled as part of his agronomic program and which, when utilized, will control one, or hopefully all, of the pests and plant diseases which are apt to afflict a given crop. This is a very large order indeed, but it is virtually certain that the research efforts will continue until this type of comprehensive combination control procedure will be available to the grower. This is already available in a few cases in the form of crop varieties with multiple forms of resistance such as the rice varieties developed by the International Rice Research Institute in the Philippines. Many would urge that more efforts be put on research moving in this direction because of the probability that outstanding combination controls via resistance and even immunity can be found and built into new crop varieties. Many of these same workers believe that a combination of breeding and cultural and biological controls may eventually allow us to do away with pesticidal chemicals altogether, or to use them only occasionally in much lower dosages.

The requirements of this new cooperative, coordinated approach have caused many to call for a completely new strategy for teaching, research and extension in this wider field, and possibly the development of a new profession built around a generalist trained to work in all of these disciplines at the field level.

Chapter 2

Agricultural Prestige Problems

LOW PRESTIGE IN AGRICULTURE

One of the principal problems not only in plant protection disciplines, but throughout the field of agriculture, is the generally low prestige given to workers in agriculture, particularly in the developing world. It is interesting to note that countries with the best agricultural programs in the developing world, such as the Republic of China (Taiwan), give considerable honor to their agricultural workers. This is also true in the developed world, in countries like Canada, the United States and Australia. In these countries agriculture enjoys considerable prestige. But even so, the prestige of the agriculturalist is considerably below that of most other professions. The professional scientist in agriculture is usually classed well below most other scientists in the prestige hierarchy. To verify this one can easily check comparative salary schedules.

This unfortunate fact raises many problems, the first being the recruitment of outstanding personnel. Luckily, in the United States and in most advanced countries of the western world, it has been possible to get enough outstanding people to do the needed research. The reason is probably that researchers have been able to split their time between the more prestigious basic research and the less prestigious applied agricultural research. Unfortunately, those who work primarily on basic research problems have often tended to divorce themselves from their counterparts in applied research.

The ignorance and negative attitudes of many writers, journalists, and others who influence public opinion concerning agriculture is also disheartening, particularly to those who believe that agriculture is the primary profession of man and that the soil is the most important of all renewable resources to preserve and improve.

Without question, agriculture should get its proper share of prestige and support from governments, particularly in countries where people are desperately poor and underfed, but this is often not the case. In the poorest countries, agriculture is usually held in exceptionally low esteem. It is most difficult to get talented people involved in the problems in agriculture, and it is almost impossible to convince administrators in these countries to spend more money on agriculture. The priorities are typically given to defense and heavy industry (Paddock 1967).

At the plant protection level, the problems are multiplied. Paddock discusses this problem primarily as it concerns plant pathologists in the developing world, but his statement applies also to the problems in all other fields of plant protection.

> The basic trouble is that while the plant pathologist may have all the facts on his side, he must deal with men who not only are not plant pathologists, but also are non-oriented towards any kind of agricultural science. A recent study of the educational backgrounds of the ministers of agriculture in Latin America showed that of the 13 whose backgrounds were known, 6 were army officers, 4 were lawyers, 2 were medical doctors, and 1 held an undergraduate degree in animal husbandry.

In the United States the top men influencing agriculture in foreign countries, especially in AID programs, often are not agriculturalists, as Paddock points out:

> A review of the backgrounds of the top 68 men in US/AID who influence the allocation of US money and the formulations of programs for Latin America shows a similar lack of agricultural training. In this group the majority are trained in law, political science, public administration, or foreign affairs with only 3 trained in a hard science.

Paddock deplores the fact that so few of the programs in Latin America have been organized to possibly succeed.

> The eminently successful program of the Rockefeller Foundation in Mexico for many years clearly merited being a model for at least a portion of the agricultural development programs in a score of countries. Yet where has any Ministry of Agriculture or US Foreign Aid Director tried to copy it?

Typically throughout the developing world, the best trained agriculturalists are not in charge of agricultural policy matters. People who do have the responsibility for agriculture are for the most part those who have been trained to handle money, or to administer large organizations, and to account for the utilization of funds in large organizations. By and large, such administrators are not only grossly ignorant of agriculture, but tend to be skeptical and fearful of technicians and scientists. They usually do not have scientific backgrounds; nor do they share or understand scientific knowledge. Consequently, they have no way of making intelligent evaluations and judgments of the various proposals they receive.

The prognosis for greater prestige for agriculture in most countries of the developing world is not good. These countries will continue to be poor until really well-trained agriculturalists are involved in the making of agricultural policy decisions and are actually holding top agricultural administrative posts. Unfortunately, the tendency of talented personnel to avoid agriculture is likely to continue in the developing world (Paddock 1967). The future actually looks even more dismal. In nearly all Latin American countries, for example, the percentage of university students studying agriculture has been declining in recent years. There were 105,000 Latin American students enrolled in universities in the United States between 1956 and 1965. Only 6 percent studied agriculture, and that percentage is now dropping. In the same decade, the students studying agriculture declined from 3 to 1 percent in Mexico, 4 to 2 percent in Panama and 2 to 1 percent in the Dominican Republic. Students have gotten the message. There is no prestige, power or even good pay in agriculture (Paddock 1967).

PRESTIGE OF BASIC VERSUS APPLIED RESEARCH

A similar problem has been equally acute. Most agricultural scientists are well aware that their prestige in the rest of the scientific community is relatively low. Further, their prestige is relatively low in the eyes of the general public, with the exception of the farm-oriented group. Because of this, and other reasons, agricultural scientists often become involved with the prestigious narrow specialties that can be supported by basic research funds, thus limiting the amount of time available for the mundane problems of applied agriculture. For some strange reason it is considered far more prestigious to work in basic or "pure" research than to be involved in research which saves a crop or doubles production.

This extreme trend toward specialization was discussed by Stakman (1964). Again, this problem extends beyond plant pathologists to the other disciplines involved in plant protection.

> I am not so concerned that plant pathology will disappear like the exploding atom. There will always be plant disease problems and crop losses from disease. What I'm concerned about is that these "specialty groups" will lose plant pathology.

Walker (1963) emphasized this same point.

> Is there danger that plant pathology might lose agriculture and that agriculture might lose plant pathology? Probably not, but they could become seriously estranged. Already certain guilds of agricultural scientists think that the term agricultural is a stigma and they want to eliminate it from their designation to

> remove the presumed taint of utility from their studies partly, so they say, because it is easier to get research grants if they are cleansed of such taints and stigmas. Surely this is a cause for concern and places us squarely before the question of motivation, of the relation between opportunity and obligation.

Agricultural scientists seeking research funds learn the hard way that it is difficult to get Federal research grants and support if the word "agriculture" appears anywhere in the grant proposal or if it is presumed in any way that the work is associated in a practical way with agriculture. The truth is that most of the groups which control research grants in the United States outside of the Department of Agriculture are dominated by scientists who are not sympathetic toward agriculture. All who are aware of the makeup of National Science Foundation panels will substantiate this point. This means that the ambitious agricultural scientist who not only has good basic research ideas but also is seriously interested in applied agriculture must, in a very real sense, become a research schizophrenic. His efforts must be split between the basic research he can get funded and the applied research which is vitally needed. Unfortunately, he often tries to keep those who are interested in his basic research from learning that he is in any way associated with the applied research phases of agriculture.

One wonders sometimes how hungry mankind has to get before administrators and policy makers realize the unique importance of preservation and conservation of soil and the development of outstanding crop improvement, management and production methods. It is interesting to note that those countries that are most affluent also have a more knowledgeable, sympathetic and helpful governmental approach toward agriculture: namely, the United States, Canada, Australia, Japan, Great Britain, West Germany, and some smaller countries such as Holland, Denmark, Israel, and the Republic of China (Taiwan). Typically, the government and people of these countries do hold agricultural scientists and farmers in somewhat higher esteem than in the developing countries which tend to completely ignore and often belittle the small farmers (Paddock 1967).

Chapter 3

Public Relations and Public Education Problems

Public relations and public education efforts in the various disciplines in plant protection have usually been ineffective where not altogether absent. A poll was taken by this author of 50 scientists and non-scientist professionals working in a non-agricultural state university in the United States. Only 3 could identify the term phytopathologist and tell what this individual did in society. These people all had advanced degrees in their own disciplines. The results certainly would have been even more dismal had the man on the street been polled.

Each of the plant protection disciplines have committees to promote good public relations. By and large, they do a poor job. Perhaps this should be expected. None of the people involved are public relations experts trained to develop good publicity. None are paid to be on the committee. Typically, they are being paid to be research scientists and/or teachers in their respective specialities. It is really little wonder that they do such a dismal job year in and year out.

The following true story illustrates the approximate level of knowledge of the general public concerning phytopathology. A young phytopathologist had just finished his Ph.D., and he was feeling very proud of himself and his profession. He attended a conference of general scientists and high school and college science teachers. When the college girl at the registration desk asked the young man's profession, he proudly replied "phytopathologist." The young lady looked up quickly and then very rapidly typed on the registration card. After filling out the card she handed him the name tag and asked how he thought Muhammad Ali would do in his next boxing match. His occupation on the tag read "Fight Pathologist."

The public knowledge of entomology is slightly higher, perhaps because most of the "bugs" are big enough to be seen and some like to live in homes. Everyone has had to fight mosquitoes and house insects, and at least the local "exterminator" understands publicity and how to advertise to the general public. Actually, in the popular mind, pest control and pest management is closely associated with the public image of the "exterminator." Many young, enthusiastic environmentalists with college educations know little or nothing about practical agricultural problems. They, too, will equate the pest control specialist with some sort of "quack" squirting poisons that endanger man's environment.

Consider the following by Shaw and Jansen (1972) which emphasizes not only the lack of public knowledge but also the enormous plant protection problem:

> The extent to which plants and animals have to be protected against damage by pests must be more effectively communicated to the general public and understood by them. Many Americans, including a segment of the scientific community, apparently do not realize that the native or natural vegetation in most areas in the United States (and elsewhere) was originally not very efficient or economical as a source of food for livestock or humans. The general public needs to understand the ecological situation in which crops and livestock are produced. These compete in a complex environment that is shared by about

30,000 species of weeds distributed throughout the world. More than 1800 of these cause serious economic losses each year. Most cultivated crops are subject to competition from about 200 weed species. From 10 to 50 different weeds infest each major food crop and these must be controlled each year. Moreover crops and livestock, not including man, are attacked by about 50,000 species of fungi that cause more than 1500 diseases. About 15,000 species of nematodes attack crop plants and more than 1500 of these cause severe damage. More than 10,000 species of insect pests add to the serious losses that occur each year in crops. In spite of the best control technology that we have been able to develop, insects, diseases, nematodes and weeds cause losses that amount to about 33 percent of potential agricultural production each year.

All agriculturalists would readily agree that the general public, in developed as well as developing countries, simply does not understand the profound implications of the foregoing quotation. If they did, the agricultural policy of nearly every country on earth would be changed quickly and dramatically.

E. Smith (1972) emphasized the great need to educate the public concerning pest control and agriculture. He stressed the importance of giving the public a general knowledge of biology and particularly that portion of biology which recognizes the importance of soil and plants to man and his future. This, by the way, is the botanical portion of general biology which tends to be neglected in primary, secondary, and even higher education. He also pointed out the importance of the public's acceptance of pest control as a legitimate part of agricultural practice and strongly felt that the public's outlook on science and technology colors governmental attitudes concerning such matters. He emphasized that effective pest control in the future simply cannot be achieved unless we begin to get a better public understanding of the problems and their importance. In fact, plant protection should become as much a part of the total concern the public has for its future as are the popularized environmental problems of the moment. Smith also charges that biologists have tended to ignore the sociobiological dimensions of their specialties and usually have not been trained to consider them creatively. There is a great need for cooperation between the sociological and biological disciplines to present and interpret the sociological implications of biological realities, particularly such agricultural realities as plant protection requirements.

Palm (1972) also discussed this problem of public education:

> We face a great responsibility in the educational field to keep the public and the legislators aware of the limitations as well as the benefits of the different methods used in pest management. Not all species of pests will respond to breeding for resistance, the use of disease producing pathogens, a sterilant, or any other particular measure of control. We must be constantly vigilant not to oversell pest management to the public as a way to free ourselves from most or all of the environmental problems attributed to certain classes of pesticides. Consequently, we need to tell through the universities and other media the story of our objectives and the methods we are trying to use to attain them, rather than leave the interpretation of results to the public alone. We need a continuing public relations program to accompany our pest management efforts since pest control is a continuing, neverending demand that requires support for ever changing requirements.

Coon and Fleet (1970) suggested the development of a new specialist trained in public education and in public extension.

> The target should not be the "public" in its broadest and amorphous sense but people in discretely recognized assemblies; and state and local government officials, civic and community action groups and the like. The salient aspect of the proposal is the development of a kind of "public educator," a hybrid of a County Extension Agent and Peace Corps worker. Not envisioned as a walking encyclopedia of the information gathered or even generated by the centers, nor as an advocate of a particular position, this person is proposed to be a catalyst.

Morison (1969) argued that we need to develop the same sort of understanding and confidence in agricultural science in the general public that now exists among farmers:

> As for less formal methods for presenting science to adults, we should devise some analogy that would do for the general public

what agricultural extension courses have done for the farmer and his wife. The average successful farmer, although he is far from being a pure scientist, has an appreciation for the way science works. Certainly, he understands it well enough to use it in his own business and to support agricultural colleges and the great state universities that grew out of them.

It would appear that the great need for better crop protection and pest management public relations and public education is true for all of agriculture. Public relations and education programs for the general urban public need to be presented in an entirely new and different way and, perhaps, through a new medium. Certainly, it will be impossible for public relations to be handled effectively by any of the separate disciplines. Only when agricultural scientists and all agriculturalists as a group realize the necessity of presenting a better coordinated public relations and education package to our urban dominated societies will agricultural science begin to get its proper hearing and, more importantly, begin to be understood and appreciated by the general public.

This effort needs to be aimed first at the formal educational system. It should begin with the first grade and go right on through the twelfth. Generally, biological and scientific programs in these grades have not been helpful or oriented to agriculture. They have been dominated by urban-oriented scientists whose primary concern in biology is the human body. This must be changed to present a well balanced approach, but it will require a concerted effort on the part of powerful agricultural groups and their allies to change it.

Public education efforts in agriculture need to be aimed at legislative bodies and at various organizations that serve as pressure groups. An equivalent to the county extension agent needs to be attached to every large urban newspaper and to every television station to present the story of agriculture in a meaningful and realistic way. In this way, the general public can come to understand the importance to mankind of the soil and of high-quality plants and animals.

The public relations for agriculture needs to present as

solid and informative a front as the American Medical Association. The layman need not understand the details of plant pathology, nematology, or entomology to appreciate the larger problems of agriculture. The petty squabbles that often exist between these various disciplines need to be buried and forgotten. All of the plant protection disciplines must unite in their public relations and public education programs and combine their efforts with those of the larger agricultural groups to present a new body of imaginative educational material for the general public. In this effort individuals trained in public education and public relations techniques, as well as in agriculture, should have the key roles. Practicing scientists have demonstrated that they are unqualified to run public relations programs. Their roles, therefore, should be as consultants, leaving the public education to those who are trained for it.

Chapter 4

Need for a New Profession

Without question there is a real need for entomologists, plant pathologists, nematologists, weed scientists, and specialists in teaching, research, and extension in these disciplines. These have proven themselves useful in a practical way and represent disciplines which must be encouraged. Most importantly, it is possible for individuals in these disciplines to find employment. Consequently, it would be most unwise to do anything to destroy what has already been built within these disciplines.

There are those, however, who feel that a new profession should be developed, and some who think it is being developed. This profession would be equivalent to the general practitioner in medicine. He would be broadly trained to work at the field level in plant protection. The role of the county agricultural agent (extension agent) has become so broad and so time consuming in many other respects, that he really cannot be expected to have the necessary knowledge or give the necessary time to plant protection problems. These have become so complex that most country agricultural agents do not feel knowledgeable enough to give accurate advice to farmers. There are, of course, conspicuous exceptions to this generalization.

In the developed world the grower has come to the point where he quite often needs an "independent broker" to give him advice in plant protection. Industrial representatives, although necessary and excellent in most regards, tend to lack complete objectivity. It is easy to understand why industrial representatives, or salemen who represent a particular company, might push one company's product in preference to another, perhaps somewhat better product, produced by a competing manufacturer. Many feel that another trained individual, one who can be more objective, is needed to represent the grower's interests. This appears to be very important in those areas of the world where large co-operative corporations and/or large multiple-family farms dominate the agricultural scene. In a few cases private consultants have been able to make a good living in this field, but this certainly is the exception at present.

In the developing world, large plantations tend to hire their own plant protection personnel. The vast numbers of small farmers in the developing world are usually ignored. They often have no one to whom they can turn for accurate scientific agricultural advice. In such situations, what would appear to be needed is a trained general practitioner sponsored by the government, an extension man trained broadly in plant protection.

Handler (1971) emphasizes the importance of developing this new profession.

> No human has yet been known to be damaged in consequence of normal usage of DDT; its untoward effects on bird and fish life appear to reflect heavy overdosage rather than proper use. In this connection may I direct your attention to a constructive suggestion, the invention of a new profession. Had we a corps of specialists trained in entomology, insect and other pest physiology and the properties and proper usage of pesticides, their attendant disadvantages could be minimized. Perhaps some safer pesticides could still be obtained over the counter as it were but society could be protected if the others could only be used by licensed, certified specialists who would know what uses were legal, adhere to maximum dosage schedules, and treat the agents

> they use with proper respect until such time as they can be replaced by suitable biological control measures. The creation of such a corps is an appropriate function of the university.

What Handler does not mention here is that even though a university can appropriately create such a corps and train them, it cannot create the demand for them. It is only as this demand grows that universities will actually move into the field and train this new corps of professionals. This training process is already beginning on a small scale, but the universities are understandably cautious about expanding into this field. It is always tragic when people are trained for jobs that do not exist.

The state of California is developing a corps called "Agricultural Pest Control Advisors" to work on particular crops. These advisors are functioning now mainly in the cotton crop.

The American Phytopathological Society is now studying the question of a new profession or discipline in this field and has actually initiated some exchange meetings between the United States and the Soviet Union concerning this possibility. Committees have been appointed and a study group has spent six to eight months in the Soviet Union studying crop protection and pest management as a broad discipline, and the possibility of developing a practical profession and/or generalist in this field (Green 1974).

No policy statement has been developed by the American Phytopathological Society on pest management as yet except that there is general agreement that pest management is not a desirable or meaningful name to most phytopathologists. There is still serious question concerning the job opportunities available for the generalist in this field. The general practitioner at the field level does not really exist as yet in plant pathology. By contrast, it has been possible for a good number of individuals from entomology to get started in this field. In all likelihood the American Phytopathological Society will continue to proceed with great caution until there is adequate evidence that there will be jobs available for general practitioners trained in a broad approach to plant protection (Green 1974).

Chapter 5

Name for a New Profession

PEST MANAGEMENT

There are many people, for the most part entomologists, who have urged that a new discipline be developed under the general name of Pest Management. This effort goes back perhaps 15 years or more and was encouraged by a group of entomologists, primarily from the west coast of the United States. Pest management teaching programs typically have been sponsored by departments of entomology and have included entomological training in addition to such topics as ecology, economics, sociology and recently, systems analysis and, perhaps, a small bit of plant breeding. Typically, however, they have not included the other plant protection areas such as plant pathology, weed science, and nematology. The pest management program at the University of Illinois is fairly typical. It is a master of science program sponsored by the Entomology Department. It includes various disciplines other than entomology but does not include the three disciplines mentioned above.

Plant pathologists and some in the other fields of plant protection have usually neither approved of the name "pest management'' nor of the more recent term, Integrated Pest Control. These are felt to be much too narrow to include the whole field of plant protection. Pest management in the past has emphasized primarily work with insects, as has integrated pest management. Historically, pest management has given only minor emphasis to the larger area in plant science that has to do with maintaining plant health in which many preventive means are utilized to prevent plant diseases, weeds and many other pests, and in which major emphases are associated with the consideration of severe abiotic plant problems.

"Pest management'' certainly is a meaningful term where insects and mites are concerned, and it can also be applied to the management of nematode populations and vertebrate pests such as rats and birds. Unfortunately, it does not concern only plants; it also includes such aspects of public health as control of household and farm pests that attack animals and humans. In the public mind it usually conjures up images of the "insect exterminator'' (Anon. 1975e), and it has unfortunate associations with the excessive use of pesticides and their adverse effects, both real and imagined.

INTEGRATED PEST MANAGEMENT

The more recent development of "integrated pest management,'' which may include plant resistance and control of insects and mites by biological and cultural control methods, although a definite improvement, is still too narrow an interpretation of the whole field of plant protection to satisfy those who are working in some other disciplines.

Integrated pest management utilizes any combination of techniques to control pests. These techniques are used in an integrated approach to keep the pest population at levels below those that cause economic injury. Integrated control depends on an adequate understanding of population dynamics of the pests, and also of the ecology—as well as the economics—of the particular crop or cropping system (Anon. 1973).

The earliest use of the term Integrated Control goes back to the work of Smith and Allen (1954) and actually the basic ideas of integrated control of insects are well over 100 years old. Many early entomologists emphasized the ecological base of economic entomology (R. Smith 1972). In integrated control, chemical control is used only when necessary, and it is used in combination with biological, cultural, and any other available methods to contain a particular pest (Smith and Hagan 1959).

Probably the standard definition for "integrated pest management" was promulgated in 1966 by FAO, United Nations, Rome (Anon. 1966):

> A pest management system that in the context of the associated environments and the population dynamics of the pest species utilizes all suitable techniques and methods in as compatible a manner as possible, and maintains the pest populations at levels below those causing economic injury.

Some evolution of the concept has occurred over the years. By 1975 breeding for resistance had been given a primary emphasis, and the use of pesticides had been de-emphasized. In general, all areas are included: weeds, nematodes, plant pathogens, insects, rodents. Also included are the following types of control: resistance, biological, cultural, autocidal, pesticides, insect attractants and repellants, growth regulators, quarantine, eradication, and others. Still missing from consideration are the abiotic problems held to be so important by agronomists, horticulturists, and plant pathologists (Glass 1975). There is still no general agreement concerning the integrated pest management concept. This point is more strongly emphasized when one realizes that the publication just cited (Glass 1975)—a product of 15 interdisciplinary consultants—was published by the Entomological Society of America but not endorsed by it.

In another study (Anon. 1975g), integrated pest management is defined as "the strategy of emphasizing the containment of pest populations by a multiplicity of methods in preference to reliance on a single method."

Here again it is emphasized that natural regulatory forces in the ecosystem are utilized where possible, and that the integrated concept and definition has grown and broadened over the years and now includes resistant crop varieties, cultural control methods, use of sterile insects, pest-specific diseases, attractants, and other biological controls, as well as the standard chemical pesticides.

With both "pest management" and "integrated pest control (or management)" the usual primary emphasis has been on the control of arthropods. In fact, Pickett and MacPhee (1965) emphasize that "integrated pest control" is a program of *arthropod population management* designed to keep pest populations below the levels of economic importance through the use of all types of environmental resistance, including plant resistance, and supplementing this by selective pesticide applications only when absolutely necessary.

Good (1974d) indicates that nematologists have utilized integrated pest management for a long time.

> I certainly feel that nematology has much to offer integrated pest management because we have perfected a greater array of control methods than any other discipline. In fact chemicals are usually used when other methods have failed. This is not to say that we cannot integrate our methods more intelligently especially with control practices for other pests.

Plant nematologists have historically approached control from the standpoint of the plant pathologist which emphasizes prevention rather than cure. Hence, they typically have used a much greater array of control methods than entomologists. Their alternatives are similar to those of plant pathologists, and until recently they have tended to use chemicals only when other methods have failed.

Also nematodes, although more difficult to count than most insects and mites, can be counted. Decisions can be made concerning the proper time to use a nematicide when the battle of control is being lost by other methods. For these reasons, the concept of integrated pest management does have considerable meaning to plant nematologists. Some plant nematologists (in North Carolina with tobacco, for example) are moving toward a systems control approach for major nematode as well as disease problems. A systems approach employs seed and field sanitation methods, crop rotation, resistant varieties, early destruction of crop residues after harvest, and use of nematicides and other pesticides where necessary. This may well be a systems or an integrated approach but it is, actually, a classical approach to plant disease control as well as to plant nematode control. If this is to be called a systems approach for integrated control, then it is what most plant pathologists and nematologists have been trying to do for many years.

This point is emphasized in the following quotation (Anon. 1972c):

> The terms "Integrated Pest Management" and "Biological Control" are frequently heard these days and plant pathologists are often asked whether these concepts have application in the field of plant disease control. The truth of the matter is that the principles of integrated disease management and biological control have been an integral part of plant pathology for decades. For the most part plant diseases are controlled by prevention not cure. In essence this means that appropriate measures are taken before the disease develops, not after an outbreak has occurred.

Although there has been a tendency on the part of some to consider "pest management" and "integrated pest management" as synonymous terms, specialists in these fields are quick to point out that this is not the case. Pest management uses all possible approaches ranging from a single component control method (such as using a pesticide) to the most sophisticated and complex controls in which many techniques are used simultaneously, as in integrated pest management. For example, not growing a crop in a high-risk area would be pest management, but it would not be integrated pest management. This means that pest management is the more general term that really applies to any form of manipulation of pest populations whether on plants, animals, or elsewhere (Anon. 1973, 1975g).

PESTOLOGY

Recently at Simon Frazer University in Canada the term "pestology" was coined. This has connotations similar to "pest management" and "integrated pest management." This program also is oriented primarily toward the control of insects and has been developed primarily by entomologists. No other educational institution or government, as far as is known at this writing, is using this term.

SUPERVISED CONTROL

Another term that has appeared in recent years was introduced by entomologists and adopted by plant pathologists under certain specific conditions. This is discussed by Chiarappa (1974). Supervised Control (essentially an entomological term) has been defined as a pest management system that relies on the limited application of pesticides under the direction of a specialist. It is based primarily on the assessment of pest densities, crop damage, and any other ecological considerations. Its goal is the most efficient and the least hazardous use of chemicals. The important point is that it is *supervised* by a specialist.

Chiarappa (1974) points out that Supervised Plant Disease Control is also possible and profitable with a few plant diseases, but thus far only with a few. He discusses the problems associated with this approach and concludes that it is profitable with such diseases as late blight of potato where the disease satisfies all the economic and biological requirements posed by a model developed by Chiarappa. Managerial factors determine whether supervised plant disease control is needed or can be justified. here again, the supervision by a specialist in such situations is essential. Supervised control programs are certainly important in specific instances, but they do not represent a very large segment of plant pathological control procedures or possibilities at the present time and it is not likely that they will.

PLANT HEALTH

In plant pathology, as H. Smith (1973) emphasizes, the equivalent of integrated pest management or control has been utilized from the beginning of the discipline. It utilizes cultural controls, plant resistance, good plant nutrition and biological control methods as standard procedures and fungicides only where necessary. In plant pathology, however, the details of the biological control phenomena are often unknown or only empirically known. Also, there is not much knowledge about economic thresholds. Until recently most plant pathology spray programs were preventive rather than curative or therapeutic systems. Now a few outstanding chemotherapeutic materials have appeared that make it possible in a few cases to cure a plant that actually has a disease.

The diagnosis of plant diseases is typically much more complex than insect infestations and usually requires a college-trained person. Consequently, it is quite difficult to train a scout for summer work to do a satisfactory field job in plant disease analysis and diagnosis.

Disease forecasting is extremely difficult in plant pathology and has only been done with reasonable accuracy for a few diseases. The knowledge of economic thresholds is essentially nonexistent since spore presence or propagule load is meaningless in plant pathology unless the plant-host population and the environment are also consistently favorable and controlled (H. Smith 1973). Smith also points out that there is no organized national plant disease survey. The national insect survey, by contrast, has been going on for some 25 years. Essentially, all of this means that even though the concepts of integrated management and control have been used constantly by plant pathologists, it is much more difficult and often impossible for them to obtain the kinds of quantitative data that entomologists have felt were necessary for making satisfactory economic judgments concerning integrated pest control of arthropods.

In plant pathology in the United States a few individuals and university plant pathology departments have stressed the term Plant Health (Smith 1974). If a school were to be devised by plant pathologists to train general practitioners, it would probably be part of a School of Plant Science and it would emphasize plant health, its achievement and maintenance. It would produce, perhaps, a Doctor of Plant Health. H. Smith (1974) emphasizes, as do most plant pathologists, that the term *pest* is negative and not inclusive, and it ignores much of what plant pathologists do. He also feels that growers tend to confuse the control of insects and diseases, and that the use of the term pest in relation to plant diseases simply aggravates this problem.

H. Smith (1976a) feels that the term Plant Health is a much more positive approach and emphasizes the important non-pest problems in growing plants that are often neglected in pest management programs, for example, plant injuries, abnormalities, weather, nutrition and soil problems, and plant breeding.

At the University of Minnesota they have recently started a Bachelor of Science degree program in Plant Health Technology. This was started in the Plant Pathology Department, and it illustrates the plant pathologists' primary concern with plant health and with the prevention of diseases and maladies rather than with cure (Anon. 1975h).

Sturgeon (1974b) has also discussed this problem as seen by plant pathologists:

> We are seeing pest management programs and so-called integrated pest management programs develop in many crop areas. These programs, for the most part, have been directed towards specific pests, in most cases insects. The terms pest and pesticide have a very negative connotation with many non-agricultural people. Special interest groups of environmentalists have expressed concern about the continued use of agricultural chemicals. Therefore, there is not only a need to improve the image and understanding of chemical usage but also a need to develop a total approach to all problems affecting plant health.

Sturgeon agrees that there is a need for a generalist, a general practioner, to be an advisor concerning plants, one who can look at a crop and analyze the problems from all aspects rather than from what he calls the "tunnel vision" of agronomy, entomology, nematology, or plant pathology. He emphasizes that this practitioner or "Plant Health Consultant" needs to be supported by all of the specific knowledge of the various specialists and also by the county extension directors, the area agronomists, entomologists, horticulturists, the chemical, technical and sales representatives and all others interested in plant health. He feels we have specialized too much and that the time has come for the general practitioner to rescue us from our excessive specialization. He recommends the term Plant Health Consultant or Doctor of Plant Health for this general practitioner, and he insists that the emphasis must be on a program of plant health and a preventive program.

The fundamental reason that plant pathologists can never accept the pest management concept of the entomologist is that the term plant disease is and should be applied to "*any deviation* from the normal growth or structure of plants that is sufficiently pronounced or permanent to produce visible symptoms or to impair quality and economic value." This is essentially Stakman and Harrar's definition for plant disease in their textbook *Principles of Plant Pathology* (Stakman and Harrar 1957). This means that plant diseases would include physiological disorders and any structural abnormalities, no matter what the cause, particularly if they are of economic value or concern. These interests include the many abiotic troubles that typically are not considered by people working in pest management. It is completely unrealistic to ask a plant pathologist or agronomist to ignore this portion of a plant health program that actually is often of much greater significance than any other portion of the program.

PLANT DISEASE MANAGEMENT

The Council of the American Phytopathological Society in 1974 endorsed the development of a new national program for 1975 in Plant Disease Management involving research, extension and regulatory activities in plant disease detection, epidemiology, plant disease loss assessment and control. This effort includes the development of a planning committee and the possibility of publishing a new journal in *Plant Disease Management* (Tammen 1974). Thus a new name is appearing on the horizon introduced by the American Phytopathological Society which appears to be complementary to the term "pest management" for entomologists. This term, however, as yet appears nowhere in the literature, as far as is known.

CROP PROTECTION

A broad, general, inclusive term is used frequently in the United States and particularly in Europe, but not often by entomologists. It includes all biotic as well as abiotic problems of plants. This is the term crop protection. Whenever plant pathologists and entomologists get together, the former seem to defend "crop protection" and the latter "pest management." The entomologists usually speak from the standpoint of pest management and integrated pest control, and the plant pathologists from the standpoint of disease prevention and plant health.

PLANT PROTECTION

Happily, on the worldwide scene a new useful term is developing which may well take the field and end the arguments. This is the term plant protection. For example, the Seventh International Plant Protection Congress was held in Moscow from August 21-27, 1975. FAO, United Nations, Rome, is now in the process of developing completely inclusive plant protection programs in cooperation with many countries in Asia. These have been developed or are being developed in Korea, Thailand, the Philippines, Sri Lanka, Indonesia, Afghanistan, India, and elsewhere. An independent country-wide plant protection organization has recently been initiated in the Republic of China (Taiwan). Plant protection is the term now utilized largely in Europe, for example at the Plant Protection Center at Wageningen in Holland (Anon. 1975e). Recently a Plant Protection Department that includes all aspects of crop protection and pest management was started in the College of Agriculture, Baghdad University, Iraq.

Other countries that are predominantly utilizing the inclusive term plant protection are Korea (Anon. 1975e), Nigeria, Poland, Hungary (Anon. 1975d), and South Africa (Van der Plank 1975). Institutes of plant protection are spread throughout Russia, and these include all of the standard crop protection and pest management subdisciplines (Anon. 1975c). The American agricultural scientists who visited the Peoples Republic of China (mainland China) in 1975 found that institutes or departments of plant protection were in all the China provinces visited. These typically included teaching, research and extension in plant diseases, insect problems, pesticides (including herbicides), biological control, weeds, and so on (Anon. 1975b). In discussing his China trip, Kelman (1975) had this to say:

> Since all departments of plant pathology and entomology are combined and weed science is often included in these departments, one can consider that the concept of Plant Protection and Pest Management as a unified area has been adopted by the Chinese.

The number of countries joining the International Plant Protection Convention is increasing according to information from FAO, United Nations, Rome. By 1974, 66 countries had signed or acceded to the convention (Anon. 1975a). Even in the United States where the arguments have waxed hottest, the term "plant protection" is used to describe 22 of the 34 combined or cooperative training programs at the bachelor's level now being taught in this country (Couch 1973).

It would seem that "plant protection" might be the logical acceptable compromise for a term to include all aspects of crop protection and (plant) pest management. It certainly appears to be the worldwide trend in plant sciences. This trend is in the developing world as well as in much of the developed world. Perhaps this name will eventually satisfy the purists in the United States among both entomologists and plant pathologists as well as those in other disciplines. As recently as 1974 Tammen reported, however, that there was no committee dealing with the issue of "plant protection" in the American Phytopathological Society (Tammen 1974). Then in 1976 an Intersociety Consortium for Plant Protection was initiated to foster interdisciplinary cooperation. The respective national societies for plant pathologists, entomologists, weed scientists, and nematologists are involved in this effort.

CROP IMPROVEMENT

Finally, we come to what is really the largest problem in attempting to find an accurate name for the combination of crop protection and (plant) pest management. The truth of the matter is this whole discipline of plant protection and all of its component parts is and should be subsidiary to crop *improvement, production,* and *management.* It is only when it is approached in this context that we can hope to develop a combined control program which will fit into the economic and agronomic (or horticultural) requirements of a crop in a given environment. This thought is well developed by Buddenhagan (1975), a man who has had much field experience in the tropics and has been involved in the agriculture of the developing world for a lifetime.

> I believe it would be useful to ask ourselves what we are really trying to do under the term "pest management." Crops have a certain *yield potential.* On this criterion new varieties are touted. Such varieties, when grown by a farmer, have a certain *actual yield.* The difference between *yield potential* and *actual yield* is where we want our pest management people to be concerned to do something immediately. This difference may be large or very small and it may be due mostly to biological populations (of insects, pathogens or weeds) or it may largely be due to a physiological disease, soil nutrient imbalance, inadequate soil fertilization, inadequate water management or bad weather conditions or other factors normally of concern to the agronomist and soils man and to the agricultural meteorologist. No "pest management" practitioner will be able to sleuth such complexities in the tropics (or elsewhere). In this view, then, it can be seen that we are really concerned with overall crop production practices and especially for food crops with crop improvement (breeding new varieties). The real need then is to sleuth the intricacies responsible for the difference between *yield potential* and *actual yield,* to design research which will lead to practices to reduce this difference, and to design new varieties which will have less difference between *yield potential* and

> *actual yield* naturally without doing anything else. I believe that "pest management" as generally discussed cannot cover this real and complex need. I believe that to develop "pest management" only as a practitioneering operation for the management of pest populations as a separate entity, separate from production and crop improvement and from basic ecological research, is to depart from the most basic need of agricultural improvement in the tropics (and elsewhere).

Buddenhagan emphasizes the important truth that we need to be working more closely and cooperatively with agronomy and horticulture and that we worry too much about being separate entities. Even though Buddenhagan's position is true, we still need an acceptable general name for this broader profession. Worldwide, the term "plant protection" seems to be developing as the international umbrella term under which we might all crawl with some hope of unanimity. For that reason it is used predominantly in this work as the inclusive term for all biotic and abiotic problems affecting plant health.

Chapter 6

Manpower Needs and Job Possibilities

The need is clear, and the jobs are available for specialists in the various sub-disciplines of plant protection. Research scientists in each of the disciplines work for government, university, and industry. Most of these have a doctorate or master's degree in their specific disciplines. Some work on basic, some on applied research, and others on both. Teachers in the various disciplines function in all of the land-grant type universities of the western world, as well as in equivalent institutions in other parts of the world. Most advanced countries have agricultural extension programs of one form or another where specialists such as entomologists, plant pathologists, weed scientists, nematologists, and the like develop ways of bringing the applied research information to the grower. Many research and development specialists are also involved in industry. Most of these also have higher degrees. Those with more generalized training function as company field representatives and salesmen who contact not only retail outlets but also growers in the field. There are also many job opportunities in chemical industries for the various discipline specialists.

There is every reason for continuing to produce such specialists for the job market. What is not so clear is just what job possibilities there are for workers who are generalists in plant protection working essentially at the field level. This group is for the most part now missing in modern agriculture, except in certain specialized crops such as cotton and citrus. However, there is a growing awareness that individuals trained at the bachelor of science level, generally along broad crop protection and pest management lines, may have a real future in modern agriculture.

Consequently, some effort is being made to start training such personnel in the hope that jobs will be forthcoming. This is particularly true in the entomological field, and also to some extent in plant nematology and weed science. Plant pathologists, in general, feel that at least a master's degree is necessary for a person to function adequately in the field in diagnosing plant disease problems. The identification of plant diseases usually requires a knowledge of microorganisms and the rather difficult methods of identifying microorganisms. These techniques are quite sophisticated compared, for example, to the techniques of field identification of most common agricultural insects, and generally require laboratory and greenhouse, in addition to field, analyses. For these reasons and others, plant pathologists are not certain whether there will be job opportunities in the foreseeable future for generalists at the field level.

Osmun (1972) has suggested the following hierarchy of people needed in pest management in the future, operating essentially at the field or near-field level, and he suggests the kind of training they probably will require:

(1) The field laborer with on-the-job training. This would include scouts hired for such summer programs as insect counts, weed reports, and possibly some plant disease reports where the diseases are severe enough to identify easily. These individuals would be trained on the job or just prior to the job to identify specific pests on a given crop.

(2) The field applicators working for large growers should be technically trained personnel, trained in the operation of

equipment and having some technical knowledge of pesticides, particularly the safe handling of pesticides. They should be trained, Osmun feels, in the use of all pesticide materials and in the legal restrictions concerning their use and the utilization of all application equipment. They could be trained through workshops, extension training programs, short courses, and the like.

(3) The supervisors of these special application personnel. These people should have more training and probably a formal certification examination. Their training would include longer extension courses and somewhat prolonged extension instruction, perhaps one to two years of total training. They would be put into leadership roles and would be expected to have considerable technical competence and ability to make judgments concerning the utilization of specific control programs at the field level.

(4) Operational or advisory personnel. These would be college degree people who also would be required to take certification examinations. In all probability most would have training at the master's degree level. Most of these would be the equivalent of present extension personnel or professional industrial personnel. It is hoped, also, that there might be some new openings for private practitioners in this field. The private practitioner trained at this level is just beginning to appear.

Osmun (1972) thinks there is a definite need for these four groups of personnel at, or very close to, the field level. The first three categories are largely missing at present except that applicators now are often available in the United States, usually working through retail pesticide sales operators.

Plant pathologists still would question how often individuals trained in these ways would be of use in the identification and control of both biotic and abiotic plant diseases. Many are skeptical of effectively utilizing individuals in the field with less than a master's degree level of training. Others would be willing to try to train individuals at the bachelor's level and see whether job openings develop and whether these individuals can function creatively at the field level.

Good (1974a) feels there will be a great many job opportunities opening when the interdisciplinary base of training in plant protection is broadened:

> Broadening the interdisciplinary base would make pest management more economically attractive to growers and to private consultants. It would provide year-round and full employment of key personnel. It would provide job opportunities for pest management students now being trained in a number of universities and it will more than likely allow hiring of area pest management specialists in states where this is not now justifiable for separate disciplines.

Again, many plant pathologists would be skeptical of this statement. Others feel it is time to train a cross-discipline generalist to operate at the field level in the hope that job possibilities would eventually be the equivalent of those available now for people trained in entomology and weed science.

Fitzsimmons (1972), an industrial spokesman for the Shell Chemical Corporation, discusses the manpower situation as it relates to the agricultural chemical industry and also to plant protection at the field level.

> Anyone familiar with field conditions in major crop areas will recognize that the average farmer or contract spray operator more frequently talks to a pesticide salesman than to his university extension entomologist or his local farm advisor. It is obvious, therefore, that the field men employed by the agricultural chemicals industry exert major influence on how pesticides are used. Nearly all of these men are graduates of agricultural colleges and many have had experience as extension specialists and farm advisors. They are familiar with local problems and resources. It is inconceivable that modern pest control practices could be carried on without them.
>
> Now, however, we are discussing far more sophisticated programs using many more specific pesticides and integration of these chemicals with each other and with a great variety of non-chemical controls. A key to making integrated control programs work is almost daily surveillance by qualified personnel of pest-parasite levels, weather, crop development, and equipment availability, not to mention consideration of various economic factors and legal permissibility.

> We do not believe it is either necessary or desirable to transfer a thousand trained men from industry to the public payroll in order to get responsible advice to farmers. There is no reservoir of trained, experienced pest control specialists. Unfortunately, applied pest control is seriously lacking at the university level, which means that there is presently a shortage of trained replacements for industry, to say nothing of the expanded future need for pest control specialists by both industry and government. So our point is simply that industry offers the only effective available manpower for the implementation of the new pest control strategies. We are ready to cooperate with all federal, state, and local agencies and experts in sound adequate efforts at such implementation.

This quotation raises the question of how people will be hired at the field level and by whom. Some insist this should be a government function. This is particularly true in developing countries where small farmers are in the majority. Others urge that this be a private consultant service paid for by large growers. Others urge that grower cooperatives pay for the necessary plant protection at the field level. Others, like Fitzsimmons, feel that the private agricultural chemical industry can develop a personnel component that is satisfactory to do this job at the field level. Still others point out that the agricultural chemical industry is not in a position to be objective about integrated pest control efforts. These would insist that the development of combination control procedures that include non-chemical means must be supervised by people who are far more objective than the average agricultural pesticide salesman or applicator whose primary concern is the sale and use of his pesticides. It is not likely that efforts to minimize the use of pesticides will be initiated by the agricultural pesticides industry or its representatives.

Despite the many arguments, one thing is crystal clear: Either the taxpayer or the grower (or both) will pay the initial costs for the additional plant protection package, and after that the consumer will pay. Certainly, the paths to be followed in the developed, as well as the developing, countries are still undefined. Probably different routes will be taken in different countries depending upon local situations and governments.

What has happened in plant protection at the field level to develop jobs for trained personnel? The oldest scouting program in the United States was initiated by the state of Arkansas on its cotton crop back in 1925 (Boyer *et al.* 1962). This was initiated primarily as an insect scouting program. Gradually more factors were included, until now it includes cultural information, weather and environmental effects, biological control of insects, survey of insects present, insecticide recommendations and, more recently, use of populations and some disease information. There are approximately 150 scouts in this program. These are, for the most part, summer personnel trained intensively for this program. Most of them are college trained, or partially trained, people who are involved in one way or another in agricultural educational programs.

Scouting programs are also in operation in California on the cotton crop and in citrus. The jobs tend to be temporary but have developed to the point where they are very important in the control program. They are concerned with insects primarily but are beginning to consider other aspects of plant protection.

Good (1974d) makes the following predictions concerning the future possibilities and personnel needs in this field. He thinks that 200 state or area graduate-level specialists in pest management will be needed in the United States in the near future. He feels that an equal number, trained at the bachelor of science level, will be needed in the chemical industry and as private consultants. If cotton boll weevil eradication programs materialize throughout the cotton belt, he thinks there would be an immediate need for at least a hundred of these people trained at the bachelor or master of science pest management specialist level. The strongest current demand for entomologists is for persons trained in pest management and other applied areas, and the preference is for those who are broadly trained. Consequently, Good feels that the prognosis for the use of personnel at the field level in pest management is excellent.

Another estimation of future manpower requirements (Anon. 1972b) is based on a supposed need for one field scout for every 2,000 acres, or one professional plant protection specialist for every 30,000 acres. If it is assumed that these are realistic needs, then the field scouts needed would be 168,047 in the United States alone. These would be specialized trainees who are high school graduates. The number of professional plant protection specialists trained at the bachelor's or master's degree levels would be 11,204.

Many would insist that these figures are completely unrealistic in most places in the United States and that the need for one field scout for every 2,000 acres, for example, is vastly exaggerated, particularly for small grains and crops of low value per acre. On the other hand these estimates might be more realistic for high value crops such as apples, citrus, and perhaps cotton. Certainly, most would feel that these estimates are very high, and in all probability such numbers of field-oriented personnel would not find it possible to find employment in the foreseeable future in the United States or elsewhere, even if they were needed.

Sherf (1973) has documented the growth in extension plant pathology. These are the people actually employed in the field from 1915 through 1972. In 1915 in the United States there was only one extension plant pathologist. By 1953 there were 48; 102 in 1969, and 150 in 1972. Ninety-five of the 150 had doctorates in 1972, five had master's degrees, and others were trained at the bachelor's level. Thirty-two states had more than one extension plant pathologist. Others had only one, and a few had none or only a part-time extension plant pathologist. Sherf's analysis shows a need for field-oriented plant pathologists that is increasing, but rather slowly.

McCallan (1969) also documented an employment trend for plant pathologists (Table 6.1).

Table 6.1 does not include those people working for the agricultural pesticide industry involved as fungicide salesmen or applicators at the field level who do not have specific formal training. Such people quite often combine all pesticides in their program and would include insecticides,

Table 6.1 Types of Employers and Employment Trends for Plant Pathologists

	Employment Trend		
Types of Employer	1953	1958	1968
State	582	611	615
Federal	187	169	170
Industrial	135	142	123
Endowed Colleges and Foundations	52	46	55
Self-employed	14	4	0

From McCallan (1969)

nematicides, and herbicides along with the fungicides. The table does show clearly, however, that there has not been a strong increase in the number of plant pathologists since 1953. The trends, as shown, are quite stable showing no marked increases or decreases with the exception of decreases in the self-employed. By contrast, another report by plant pathologists (Anon. 1968) estimates that at least 50 percent more plant pathologists will be needed in 1980 than in 1966.

Cox (1976) documents the difficulties encountered by the private practitioner in plant pathology, particularly in the area of prestige and recognition by other plant pathologists. He indicates that some 50 individuals were involved in private practice in the area of plant diseases in 1974, and were recorded in the 1974 APS Membership Directory by H. Smith (1976b).

The foregoing rather erratic data doubtless account for the extreme caution shown by land-grant universities in the United States concerning the development of generalized training programs in plant pathology. It is certainly not clear to plant pathologists just how the generalist will be trained and just how effective he can become insofar as field problems are concerned.

By contrast, in entomology the picture is becoming much clearer, as indicated by a recent message from one of the past presidents of the Entomological Society of America (Adkisson 1974). In an article entitled "Opportunities for Professional Entomologists in Private Practice," he emphasized that there is a new possibility for the development of private consultants and private enterprise for entomologists. For the first time during his professional career, the per acre cost to the producer for employing a professional entomologist to handle pest problems was less than that of one or two needlessly applied applications of a pesticide. He indicated that the combined pressures on the grower—of shortages, high prices and hard-to-control pesticide-resistant insect strains—has created a good opportunity for the professional entomologist who wishes to enter private practice on several crops. He lists cotton, grain sorghum, soybeans, peanuts, fruits and vegetables, and some other crops that have high pesticide requirements. He showed that Texas A & M University has recently lost seven entomologists to private entomological practice and that all are doing well financially as private consultants. He mentioned also that the Director of the Nebraska Agricultural Extension Service recently resigned to become a private entomological consultant. He also suggested that, at present, the way to reduce pest control costs for many farmers is to hire a consultant. Certainly, when the above is also true in other plant protection disciplines, private consultants will appear rapidly, and the broadly trained generalist will be needed.

Most countries in the developing world have very few extension people working in plant protection at the field level. Some have none. Where they do exist, most are hired by the government. Occasionally one will find a representative of an agricultural chemical company functioning at the field or near-field level but, in general, these personnel also are in very short supply in the developing world.

The program, reported by Chiu and Yen (1972) in the Republic of China (Taiwan), is particularly outstanding and might hopefully become the pattern for those developing countries where the small farmer is typical. In 1972, there were 1,208 persons engaged in full- or part-time employment in plant protection (crop protection and pest management) in Taiwan. Of these persons 73.1 percent were graduates of agricultural high schools, and these were extension workers of one sort or another hired by the government. Teaching, planning, or research activities involved 26.9 percent, a total of 323 people. There were 31 holders of doctorates, 26 masters of science degrees, and 137 bachelor of science degrees, and the balance were high school graduates. Some of the latter group had considerable training but had not yet completed the bachelor's requirements.

When one considers the size of Taiwan and its agricultural acreage, as well as the number of people living there, this is a very impressive number of people working in plant protection. Another notable feature in the plant protection picture of Taiwan has been the availability of a large number of extension workers at the field level where they perform a surprising range of activities.

The agricultural extension system consists of organizations knows as Farmers' Associations. These associations occur at the township, county, and the provincial levels. Most of the 885 plant protection extension workers are attached to different levels of Farmers' Associations so that they are in constant contact with farmers' needs. The remainder are employed by township government offices. The upper-level personnel are hired by various teaching, research, and extension organizations, and institutions in the country. However, in recent years in Taiwan there is beginning to be a small core of private consultants working in plant protection. This outstanding program of plant protection needs to be studied very carefully by individuals interested in developing equivalent programs in other developing countries where small farmers predominate.

Chapter 7

Legislation: Governmental Regulation of Pesticides and Certification of Applicators

GOVERNMENTAL PESTICIDE REGULATION

Because of the excessive use of pesticides—particularly insecticides, there has been a movement in recent years not only by environmental groups but also by concerned agricultural groups, and even by the agricultural chemical industry itself, to control more closely through legal means the use of pesticides that may be potentially damaging to man and other warm-blooded animals, and accumulate in the environment.

Actually, legal and statutory restraints were built into the first pesticide toxicity and residue laws in the United States going back to the middle 1930's. Each of the early laws developed to control tolerances and residues was actually initiated by an agricultural group or by the agricultural chemical industry. Until recently these laws seemed adequate. In fact, it was not realized that a few pesticides were actually accumulating in the environment and possibly did not degrade rapidly enough to be environmentally safe for continued long-term use. This was particularly true of such materials as DDT and some of the other chlorinated hydrocarbons.

Certainly, any necessary legal or statutory restraints on the use of pesticides should be developed cooperatively by agriculturalists, the agricultural pesticide industry, appropriate government bureaus, and concerned environmental groups. Without question, real problems have arisen and these should be solved by careful research, paying particular attention to residue, denaturation, and toxicity problems, most of which are extremely complex.

Sometimes pesticides have been misused by farmers and others who have felt that if a little bit is good then more must be better. There has been much carelessness in following the directions given by the agricultural pesticide industry. This undoubtedly has caused many of the problems that have developed. This is also true in portions of the developing world where poorly trained people have often utilized pesticides indiscriminately, without reading labels, and without actually knowing what it is they are trying to control or at what dosages control is possible.

Another important problem was outlined by the Jensen Commission in 1969 (Anon. 1969; Carlson and Castle 1972). This commission concluded that the public demand for attractiveness, particularly of vegetables and fruits, has stimulated excessive use of pesticides. This essentially cosmetic use of pesticides to achieve better control than is actually necessary is a major problem. The Jensen Commission recommended a downward revision of federal grades and federal quality and marketing standards of fruits and vegetables. This would mean that consumers would have to share the cost of using fewer insecticides by accepting a somewhat lower standard or quality of insect and plant disease control. E. Smith (1972) emphasized this point also.

> We need to reassess market standards (of quality). Some of them impose exacting specifications that have no significance in terms of nutrition, sanitation, or aesthetic value. On what basis are standards established for the number of insect fragments that constitute contamination in a food product?

The discriminating tastes of the modern consumer in the western world is a serious problem and a massive educational program is necessary to convince consumers to accept products of equal nutritive value but of somewhat less cosmetic attractiveness.

Carlson and Castle (1972) think that pest control programs should be developed and altered by government regulations as follows:

(1) Acreage allotments could be transferred to areas where pest densities are quite low.

(2) The patent life of the narrow spectrum chemicals that are difficult to manufacture profitably should be increased.

(3) The application of chemicals, particularly those dangerous to man and warm-blooded animals, should be restricted to licensed personnel.

(4) Acceptable vegetable and fruit grades should be lowered. This would have the effect of reducing the amount of pesticides used to achieve adequate control.

(5) Smaller quantities of certain pesticides that accumulate in the environment, such as DDT, could be approved and sold to licensed applicators only. These, then, would be used under strictly controlled conditions and only when absolutely necessary.

CERTIFICATION OF APPLICATORS

In recent years there has been a gradual movement toward the certification of individuals working with pesticides in plant protection. The United States Department of Agriculture is working with states to develop certification programs for private plant protection specialists. It is felt that certification will help assure farmers not only of the experience but also of the ability of such specialists to employ restricted pesticides in a safe and effective manner (Anon. 1972b). Several states have recently adopted certification programs. Mississippi now requires licensing for entomological, plant pathological, and weed control consultants. Certification is also required in both California and Washington. Oral and written examinations are required and a fidelity bond must be posted. In California, the law reads:

> No persons shall advertise in any manner to render professional services or solicit business as entomological, plant pathological and weed control consultants without first obtaining a license.

Plant pathologists have tended to lag in the registration of professional plant pathologists, perhaps because only 17 states have plant pathologists on their regulatory staff (H. Smith 1974). In fact, at the 1974 Vancouver, British Columbia, annual meeting of the American Phytopathological Society a proposed registry of professional plant pathologists, which would have been a form of certification, was voted down by a narrow margin. Some plant pathologists are vitally interested in the development of certification programs particularly for field-oriented and extension plant pathologists. Others object to certification and do not feel that it is an important matter (H. Smith 1974).

The Entomological Society of America has sponsored an American registry of certified entomologists. This is a nonprofit activity and functions through the Committee on Professional Training Standards and Status. This certification of entomologists by their professional organization goes back to December 3, 1970. Actually any professional entomologist in good standing as of that date was given a certification classification category based upon certain criteria developed by the Entomological Society of America.

In the *Federal Register* of February 22, 1974, the administrator of the Environmental Protection Agency (EPA) proposed regulations that would establish standards of competence for *applicators* using "restricted" pesticides. This proposal is still being considered. Any person interested was able to file comments on the proposal prior to March 25, 1974. The EPA will eventually be responsible for certification of personnel to handle pesticides. Each state is to develop its own guidelines for both certification and training of pesticide applicators. The deadline for the development of these certification and training programs was October 1, 1976. These state programs must be approved by the EPA based upon its Certification of Pesticide Applicators regulations in the *Federal Register* for October 9, 1974 (MacKenzie 1975). In particular, certification will be required of those persons allowed to use chemicals classified as "restricted" because of environmental hazard, and persistence or toxicity to men and animals. The EPA will establish standards of competence for pesticide applicators. These standards of competence will not be as necessary for "private applicators" as for "commercial applicators." No certification will be required for the large number of reasonably safe materials classified as "general use" chemicals.

The move toward certification of personnel is definitely in progress and probably is necessary. It certainly will be necessary to have a license to utilize many of the more dangerous pesticides in the foreseeable future. Some pesticides undoubtedly will have to be applied only by trained professional applicators. Other pesticides probably will be made available to individuals on prescription basis as is now typical in the human medical profession.

Statements coming from appropriate United States government offices indicate that the federal government is still developing appropriate regulations and laws on all aspects of certification and pesticide safety and that this matter is essentially still in a formative stage. There is at present a heavy entomological emphasis and bias in most regulations (H. Smith 1974).

Of extreme importance is the adequate training of competent personnel; control of pesticides cannot come just through new laws (Anon. 1972a). The use of pesticides is tied inseparably to the competence of the user, and any laws must be accompanied by an adequate educational program to produce enough well-trained personnel to control, via prescription techniques, the movement and use of dangerous pesticides. The following quotation emphasizes this need (Anon. 1972a).

> It is clear that all of us will soon be operating under the shadow of national legislation that requires state certification of personnel and provides stringent restriction on certain use of certain

> pesticides. We can classify pesticides and pass laws until we are all but strangled, but none of this will work unless we have a sound and reasonably uniform program of training across the country. Even if industry can develop new chemicals, it is a truism that a whole cache of diamond-studded pesticides would not be worth a bag of dust without trained personnel to use them.

H. Smith (1974) indicates that his U.S.D.A. agency is putting much of its effort into training pesticide applicators (in plant pathology) for future certification and/or licensing.

Two areas are still very much in a state of flux. There are, for example, no pesticide disposal mandates in the proposed federal regulations at this time. This is potentially a very important problem. Interim policies on many of these matters are being left up to the states (MacKenzie 1975). Also it is clear that the worker protection regulations are in limbo at present. For example, Richard M. Monk, safety engineer of the Division of Health, Office of Standards Development, United States Department of Labor (Monk 1974), states that the Occupational Safety and Health Administration (OSHA) has *no* relevant regulations that apply to pest management and crop protection. The OSHA emergency temporary standard for field worker protection was vacated by the courts and has not been repromulgated. But on May 10, 1974, the Environmental Protection Agency promulgated a standard on field worker protection that covers essentially the same issues as the emergency temporary standard of the Department of Labor.

Other industrialized countries vary in their pesticide laws and certification requirements, but most are approaching the problems in ways that are somewhat similar to those being utilized in the United States. And they seem to be having somewhat similar problems.

REGULATION IN THE DEVELOPING WORLD

On the world scene and particularly for the developing world the Food and Agriculture Organization (FAO) of the United Nations, Rome, has sponsored several conferences and working groups that have examined possible legislation and the problems of certification. In 1969 a publication, "The Guidelines for Legislation Concerning the Registration for Sale and Marketing of Pesticides," was published jointly by the FAO and the World Health Organization (WHO) of the United Nations. In 1970 FAO published "Model Scheme for the Establishment of National Organizations for the Official Control of Pesticides." This was a result of the efforts of an FAO working party of experts on the official control of pesticides (Section A on Legislation). In 1971 FAO published Agricultural Development Paper No. 93 entitled "Manual on the Use of FAO's Specifications for Plant Protection Products." This also was the result of an FAO working party of experts on the official control of pesticides (Section B on Specifications).

Most of the countries of the developing world have few or no restrictions on the use of pesticides at this time although there are some countries which are considering regulations or are in early stages of developing them. The Republic of China (Taiwan) is one country that has developed a rather sophisticated body of regulations for the use and control of pesticides. Eventually all countries will be forced to face up to the necessity of controlling the more dangerous pesticides and making sure that these are utilized safely by adequately trained personnel. Much work is needed at the national and international levels on the development of sensible laws for the use of pesticides and practical programs of certification and licensing of applicators and other workers. Much concerted effort on the part of professional bodies will be necessary over the coming years to cope with these very important and complex problems.

Chapter 8

Cooperative Efforts: Problems and Needs

The plant protection (crop protection and [plant] pest management) subdisciplines are fragmented in the United States and have developed independently. They include entomology, plant pathology, nematology, weed science, portions of zoology, agronomy, and horticulture, as well as portions of several related disciplines, particularly those which concern the many abiotic maladies of plants. This fragmentation was a natural growth of the several disciplines and is related closely to the nature of the human mind and its need to fragment and analyze nature so that it can, hopefully, later be synthesized and understood. However, at the field level in agriculture, nature is a welter of interrelated organisms, physical structures and forces in a constant interplay that "boggles" the mind. Hence, to the grower at the field level, nature does not appear to be fragmented but appears to be a unified piece. This undoubtedly accounts for the farmer's frequent annoyance with plant science specialists who insist on discussing only their own small discipline when they discuss his problems in the field. It also explains the frequent annoyance of the generalist with the specialist.

In recent years there have been formal efforts to combine the plant protection disciplines in various ways. Prior to that time many informal relationships developed over the years between people working in the different disciplines in plant protection. These informal cooperative working arrangements have been very effective in some parts of the world and particularly so at the field level in some states in this country where an informal team approach to field problems has been the rule rather than the exception. Actually, in some states a team representing the different disciplines routinely travels to examine fields where crops are suffering from various crop pests and maladies. In passing, it should be stressed that this simple system may be the cheapest, most efficient and effective method of initiating meaningful discussions concerning needed cooperative research or control efforts.

Some feel that there has been an enormous amount of progress made in the recent development of formal cooperative efforts among plant protection disciplines. Others think little cooperation has occurred. These contrasting views are presented in the next two quotations. Good (1974d) has this to say:

> Pest management is receiving great attention in all professional societies in the United States, The American Entomological Society, Society of Nematologists, Weed Science Society of America and the American Phytopathological Society. Greater progress is being made than I anticipated 2 years ago in merging the disciplines into one crop protection philosophy under pest management. As a result of the pilot-demonstration programs financed by the U.S. Department of Agriculture, there is an increasing awareness of the need for a complete crop protection service if pest management is to become practical and economical.

By contrast, Furtick (1975), an outstanding weed scientist, says the following:

> Although the interaction between disciplines has been talked about a lot in relation to integrated pest control, in fact the cooperation has so far been minimal.

Apple (1974) discussed recent efforts toward cooperation.

> We have made considerable progress in recent years in effecting integration across crop protection disciplines, but the lack of interdisciplinary and inter-departmental communication and interaction remains as a principal deterrent to agro-ecosystem management. We can no longer permit each disciplinary group to pursue independently their research without regard to the impact of their actions on other components of the productive system. We are just beginning to recognize the negative interactions involving fungicides/insects, insecticides/disease organisms, agronomic practices/pest problems, and so forth. Experiment Station Directors can do much to stimulate this type of integrated research through their project development and funding practices.

Good (1975) made these additional remarks concerning cooperative efforts between the disciplines:

> To date entomologist and weed specialists are cooperating to a greater extent than other specialists in developing integrated pest management programs. I am afraid it is a lack of conviction and commitment on the part of plant pathologists and nematologists that we are not involved more heavily.

Many would not agree with this statement but since Good is a nematologist we, at least, cannot berate him for flagellation of disciplines other than his own.

The real need for cooperative effort has been recognized by thoughtful leaders in all of the disciplines for many years. For example, the following statement by Painter (1951), an outstanding entomologist and certainly one of the foremost proponents of breeding plants for resistance to insects, dates back to his earlier cooperative efforts carried out long before 1951.

> In plant breeding the pursuit of any one genetic character such as resistance, without continual attention to others may quickly make a strain of little use. So many facets are involved that there is an important requirement for team research. Formal written agreements can facilitate such team work and may be administrative necessities, but real success depends on sharing work as well as credit, and on the true meeting of minds in the field plot, laboratory, or greenhouse as well as about the conference table. The workers require a broad biological background with particularly an understanding of the problems and possibilities in the other's field of work.

Plant pathologists, probably because many have been so closely involved with cooperative plant breeding efforts, have been less enamored than others by recent suggestions to combine disciplines. Also, they have been under less pressure than other disciplines from environmental groups concerning the use of pesticides. However, in recent years a realization by some of the most thoughtful and farsighted plant pathologists has developed which points to a movement in plant pathology toward more formal cooperative efforts, largely owing to economic necessity. This is illustrated by the statement of Sherf (1973), a respected plant pathologist with long-term experience in State Agricultural Extension Service at the field level.

> Recommendations for agriculture today can evolve only as the result of group effort involving plant pathology, entomology, agronomy, horticulture, agricultural engineering, and agricultural economics. No longer does the plant disease specialist work alone nor can he for long consider only diseases when he deals with the most efficient farm technologies of the 1970's and 1980's. Only knowledge developed by a team of specialists looking at all aspects of a new farm practice, making compromises when necessary, and with necessary field testing, can be utilized by today's farmers. Our audience is no longer the farmer alone. Rather, it involves society and other members of the agrobusiness partnership, for example, the banker, seedsmen, farm implement and chemical dealer, the U.S. Department of Agriculture personnel, and others. Total recommendations come harder now than formerly since farming is more competitive, more costly, and provides a narrower margin for error. Thus recommendations must be based only on careful research.

The following quotation (Anon. 1968) emphasizes an important consideration to plant pathologists that is seldom of great importance to most others working in plant protection, namely, that much of plant pathology concerns abiotic maladies of plants, and these abiotic maladies are becoming more important, requiring more cooperative research with other disciplines.

> In the future plant pathologists must work with scientists in related disciplines to learn more about the nature and effects of the many noninfectious agents that cause plant diseases and, in

> particular, to develop a better understanding of the influence of abiotic factors on the development of pathogenic diseases. It is obvious that the amount of noninfectious damage to plants will increase with increases in population, intensification of agriculture and industry, and the use and reuse of available water supplies.

The cooperative efforts needed here are usually with disciplines other than the usual plant protection disciplines (for example, with plant physiology and biochemistry).

Officially the American Phytopathological Society is still holding back and is extremely cautious in any efforts toward the development of formal cooperative efforts with other disciplines. Many plant pathologists in correspondence present a common theme, essentially that at present little is to be gained by formal cooperation and much can be lost. They all cite the many cooperative efforts with other disciplines now in progress, both formal and informal, and particularly emphasize that much of the cooperation occurring now is with non-crop protection non-pest management disciplines.

However, in 1976, according to Bruehl (1976), the American Phytopathological Society has urged the development of and joined the Intersociety Consortium for Plant Protection (ISCPP) (Anon. 1976) which also includes the Entomological Society of America, the Society of Nematologists, and the Weed Science Society of America. However, Bruehl indicates that the Plant Pathologists thus far are the only group planning to meet with the International Congress of Plant Protection in Washington, D.C., in 1979.

Among nematologists, perhaps, the position of Good (1974d) concerning the need for cooperation is fairly representative:

> Because of the frequent interdependence between nematodes and other pests, in the future, management of nematode populations must be considered in a broader framework of integrated management of all pests. Too often recommendations for the control of a specific pest, regardless of the method employed, are made without consideration for the management of other pest problems of local or area-wide importance. Because present pest management practices seldom consider all components of pest control, important interactions are frequently overlooked. These

interactions can be detrimental or beneficial. When all components of pest management and the agro-ecosystem are considered, many of the detrimental interactions could be avoided or minimized while the beneficial interactions could be accentuated.

When cooperative efforts are discussed among plant protection personnel one seldom hears the point of view of the agronomist and horticulturalist. Yet that view should be expressed and must be accepted in the long run. Plant protection, whether we in these disciplines like it or not, is just a part—and a relatively small part—of overall crop improvement, production, and management. It invariably must be subservient not only to the agronomic and horticultural needs of a crop but also to the long-term conservation needs of a given soil and the economic requirements of the crop. These factors dominate the thinking of the grower and rightly so. Plant protection is considered by the grower only when absolutely necessary. If personnel working in plant protection would only realize and admit where their rightful place lies in agriculture, they might be much more receptive to the notion of sincere cooperative interdisciplinary efforts to achieve the necessary goals of plant disease and pest control within the larger framework of crop improvement and production.

The above point of view was also expressed at the Third FAO Pest Management Session in 1971 (Anon. 1971).

At the research and development stages there is, therefore, a special need for collaboration between scientists in the different disciplines. For example, collaboration with agronomists and soil chemists is essential where cultural practices and soil fertility affect pest incidence. Furthermore, since the basis for most forms of plant resistance to insects is often so sophisticated, plant breeding for resistance may need the collaboration of plant breeders, entomologists, behaviorists and physiologists. Before introducing new crop species or varieties to a specific agro-ecosystem their response to prevailing pest complexes should be determined as carefully as their response to climatic and soil conditions. Weed control with herbicides has profoundly affected the ecology of many agro-ecosystems, but modern selective herbicides can now be used to adjust the weed species' complex as well as the overall abundance of weeds. That collaboration with weed control specialists may prove valuable

> in integrated pest control programs is shown by the evidence that different weed species can variously act as a camouflage against colonizing insects, as a vital link in the maintenance of natural enemies, or conversely, as an undesirable source of the pest. In addition, the use of fungicides or other pest control chemicals by plant pathologists may greatly interfere with or eliminate the activity of natural enemies keeping a group of key insect pests under control or influence the rate of increase of pest populations.

This last statement not only emphasizes the importance of cooperative efforts but also emphasizes the many conflicts in recommendations that do occur at the field level.

In the long run there are many advantages to the grower and for the future of agriculture if cooperative efforts are encouraged and achieved. R. Smith (1972) mentions three advantages of coordinated control efforts:

(1) The economies of combining the resources of several agencies and several disciplines to solve a single problem.

(2) The use of the area concept of pest management rather than pest management on a specific farm, which also quite often introduces very important economies.

(3) The fact that the direct beneficiary, the grower, is going to be making the major investment and paying the bill. This will also encourage cooperative efforts because of his desire to pay the smallest bill possible.

Wilcke (1972) presented six problems that need to be approached cooperatively. He feels they are too big for any one discipline to attack alone.

(1) The details concerning the biologies of the various pests as well as beneficial arthropods and their many relationships to the environment.

(2) The development of reliable methods of estimating population levels of organisms.

(3) The development of reliable methods of estimating economic threshold levels of pests and other organisms, where possible.

(4) The development of methods of accurate prediction of

increase and decline of pest populations as influenced by beneficial organisms, other methods of biological control and abiotic factors, like weather, for example.

(5) The development of detailed knowledge concerning the various influences affecting pests and beneficial organisms by the use of pesticides on a particular crop.

(6) The many efforts, sometimes conflicting, of the use of varying cultural practices, irrigation practices, harvesting procedures, cover crops and so forth as they relate to a given crop.

All of these are exceedingly complex problems that must be examined by a variety of disciplines from their own biased point of view and then these data must be combined to develop a synthesizing body of data to help resolve conflicting control recommendations, conflicting agronomic recommendations, and the like.

Certainly, the economic necessities of agriculture will gradually drive the plant protection disciplines together into cooperative efforts whether they choose voluntarily to do so or not. The frequent conflicts in recommendations eventually will mean the development of cooperative efforts to reconcile these conflicts. Not only the economic but the agronomic and environmental facts of life will drive the disciplines together into more cooperative efforts.

In the developing world, probably because of the influence of FAO programs and other international influences that have tended to bring the plant protection disciplines together, the problem of developing formal cooperative efforts will not be as great. The development of the cooperative Plant Protection centers sponsored by FAO (UNDP) in many countries in Asia which are now in progress, certainly will stimulate complete cooperative research and extension efforts.

Chapter 9

The Human Problem or Just "Human Cussedness"

Whenever the effort is made to achieve close formal cooperative effort between disciplines, the collaboration often seems to break down somewhere. The real reason for the breakdown is likely to be what we might call "human cussedness" or plain contrariness. This endemic trait of human nature seems to be as prevalent among scientists as elsewhere in the human race. Perhaps the very requirements of accurate scientific inquiry attract the individual who tends to be a strong-willed "loner" and one who is more exacting, demanding, persistent, jealous of perquisites, and the like. Whatever the reasons, the problems of formal cooperation in science and among scientists are considerable. Problems of politics, power, credit, and control usually seem to rear their ugly heads. Quite often a formal agreement for cooperation collapses simply because individuals involved will not cooperate with each other despite the agreement. In other cases an informal cooperative arrangement will work very nicely because the individuals involved sincerely believe in the importance of the cooperative effort and are willing to sacrifice personal differences to achieve their common goal.

The "prima donna" or "golden boy" syndrome is particularly acute in science. Each individual researcher frequently seems to be attempting to build his own little kingdom, a group of associates who operate under his influence. Granting agencies, particularly those aimed at basic research, tend to foster this "prima donna" syndrome. Certainly, there is some justification for this. We can all point to the fact that very few great basic scientific discoveries have been made by way of team research or a committee. Most of the truly "breakthrough" discoveries have been made by one man or perhaps by two or three men working in usually informal cooperative teams.

Anyone who has attempted to develop or has been involved with a relatively large-scale cooperative research project that includes several disciplines will be aware of the acute problems associated with politics, power, pay, perquisites, and especially control of the projects. The following response is a classical anonymous example of the problems that arise. This particular case also would certainly have to be classified as one of the more creative efforts in formal cooperative research in plant protection to date in the United States.

> If I understand your letter correctly you would like an evaluation of the cooperative aspects of our project. We have had our administrative problems but not because of any basic conflicts between the disciplines, at least not here. Our major problem has been a poorly conceived administrative structure combined with the lack of effective leadership. We have just recently taken steps to correct the situation and I think our project has been improved tremendously as a result. The Table of Organization with which we muddled along for two years was a basic problem. The actual work in the field was carried out by the manager, the trainees, and the scouts. This part of the organization functioned well and remains unchanged in the new administrative structure. The manager was supervised by a committee of four, a mistake which nearly caused the collapse of the project. Our original plan had been to have a director who would be advised by the committee but we were not able to hire the caliber of person we wanted on a short term basis. So we settled for a lower level managerial position supervised by the committee, as I said previously, a bad mistake. The idea was that

the four member advisory committee would be the policy making body for the project, while the manager would be responsible for the day to day operations. What, in fact, happened was that the advisory committee could not always agree on policy and could not always get together when some of the necessary decisions had to be made. They represented different disciplines and their opinions varied. The Chairman of the advisory committee, who was a very strong minded individual, took advantage of the opportunity to assume control (not leadership) of the entire project. When the advisory committee could not agree or could not meet, its chairman made the decisions. Gradually he took over the day to day management of the operation, relegating the Manager to a lackey position and treating the other members of his committee as advisors. In writing the annual report he, in fact, named himself "Acting Director."

The input from the cooperating scientists took place in large meetings involving the four members of the advisory committee and 8 to 10 of the other scientists. These meetings can be generally characterized as lengthy, heated, and unproductive. There were too many people and too many issues involved in one meeting to nail things down and make progress.

The Steering Committee has served essentially a figurehead function. They have met a couple of times to consider the loftier objectives but it had little direct input into the project. This committee, however, is essential to the funding of the project and it appears unchanged in the new Table of Organization.

The need to set up an administrative structure that included other crops gave us an opportunity to reorganize our project and dislodge the "Acting Director" who had generated considerable hard feelings and discouraged many of the cooperators to the point of dropping out of the project. Under the new organization, a new Steering Committee was formed to oversee all of the projects with which we are involved. Each project has its Coordinator, who in the case of our project happens to be the Chairman of the Department. The Coordinator in my opinion is the key person in the organization. The ultimate responsibility and authority (both on paper and in practice) must be vested in one person rather than in a committee. It matters little whether the Coordinator is an entomologist, plant pathologist or whatever by training. The important thing is that he is capable of effective leadership. This, of course, means that he must have the respect of the people participating in the project, that he must be open-minded and show no bias to a particular discipline, and that he must be committed to the overall goals. Without good leadership in an organization of any size there's bound to be someone who will seize control and that person may or may not be a good leader, as we so sadly learned.

In the new organization the various cooperators are organized into committees by disciplines, each chaired by the respective extension person in that discipline. This breaks the large body of cooperators down into workable groups that can meet and work out the details pertaining to their own fields of interest. The groups are in communication with one another but have an input to the project itself only through the Coordinator. The Coordinator irons out any conflicts between the disciplines, which for us works considerably better than everyone shouting at one another in one large meeting particularly where the entomologists outnumber everyone else by at least 2 to 1.

One other important detail in the new organization is the development of an Advisory Committee made up of growers, extension agents, and representatives of the agricultural chemical industry. We learned from sad experience, that not only do these people have a lot to contribute but they also do not like being left out. Their cooperation is essential to the success of the coordinated project and their cooperation is ensured only if they feel that they have some input which is needed.

This is about all I have to say. It is given only as an example of the kind of problems that can arise in a large cooperative project such as this and some of the approaches toward solutions of these problems. In looking over this letter there is one comment I would like to make concerning the problems of cooperation. An entomologist has always been the principal figure and Coordinator in this project and it is my impression that it would be exceedingly difficult to get the approval of entomologists involved for anyone but an entomologist to be the Director in charge.

Another anonymous comment concerning another large cooperative project emphasizes similar human problems.

Although we have never faced a real struggle for leadership on a discipline basis, in my opinion it would be almost unthinkable for the entomologists to accept a plant pathologist in this role. I have suggested a couple of plant pathologists whom we might consider for the key position and always felt the entomologists wouldn't have considered this prospect seriously.

Some of the plant pathologists participating on the project don't feel we have been given equal treatment. We have detected the tendency of the entomologists to monopolize the technicians

> and dictate the program, but we made it very clear from the beginning that this would not be tolerated. Consequently, we have not been in the back seat too much.
>
> Certainly our entomologists recognize the importance of participation by plant pathologists and other disciplines. They feel strongly that every time a grower has to spray fungicides there is real danger that an insecticide will be added for good measure. Thus, they recognize that their chances of reducing insecticide applications is somewhat dependent upon our ability to reduce fungicide applications.

These very revealing comments from scientists involved in large-scale cooperative research efforts emphasize that it is seldom the scientific question that cannot be handled in a cooperative manner. Rather, it is the human problems that cause the efforts to bog down and often grind to a halt. These problems include who gets the credit, whose names appear on the publications and in what order, who is listed as Director, who gets the new equipment, in whose laboratory will the new equipment be placed, who gets the secretarial help, laboratory assistants, field helpers, who among the cooperators has to accept a footnote in the publications, or perhaps get no official credit at all. In short, the solution of these problems of "human cussedness" are usually the keys to the success or failure of the project.

The many personnel problems can only be solved by fair agreements and proper organization, by either giving no official credit to anyone or essentially equal credit to all in the cooperative venture, with the possible exception of the director. It is most helpful, too, if the director is not only competent and strong-willed, but also unselfish by nature—a "rare bird," by the way.

One of the principal problems is centered in the way research and extension funds are obtained and administered. Most basic research funds are given to a single man as the director of a laboratory or a project; and this has meant the development of an extremely competitive effort between the various "prima donnas." Each tries to outmaneuver other individuals for the inevitably limited funds. Such individually controlled projects typically are not aimed at the solution of large problems but usually are built around the consideration of narrow basic research questions. It is very difficult to get support for a cooperative or coordinted research project that is aimed at the solution of a group of related complex problems. Recently, there has been some correction on the part of granting agencies which have finally recognized some of the problems associated with "prima donna" research, one of which is that the big practical problems don't tend to get funded, researched, or solved. In recent years the International Biological Program and some other large equivalent programs in the United States and elsewhere have been designed to consider the solution of large complex practical problems. These have helped somewhat in the development of a more cooperative approach toward large-scale problems that are essentially interdisciplinary.

Another type of useful grant, which is often difficult to obtain, is the block grant that is given to an institution for no designated, specific purpose. Such grants have been given to land-grant universities by the Department of Agriculture for many years, and without question these have been administered with the least red tape of any of the grants available to agricultural research workers. In these block grants the problems are determined at the local or state level and the money is distributed by local research directors to individuals or groups involved with the solution of these practical state or local problems. Block grants can be utilized quite easily in the development of cooperative research projects. They also give an opportunity to local or state administrators to face up to the real applied research problems that occur in a given location.

In the future, much more success in solving practical complex interdisciplinary problems can be expected if in the development of cooperative research projects the granting trend is toward either large cooperative research efforts like the International Biological Program or block grants given to specific institutions. This would be with the understanding that the money will be utilized in an attempt to solve serious practical local or regional problems.

This is not to suggest that all support for individual research aimed primarily at basic problems should be ignored or eliminated. Actually, these efforts should be increased. What is important is that both types of research projects be stimulated in all ways possible so that not only the necessary basic research can be supported but also the important interdisciplinary research, both basic and applied.

Somehow, too, administrators must develop a technique for paying and rewarding those involved with interdisciplinary, cooperative research programs in better ways so that their personal sacrifices of prestige, power, control and other perquisites are acceptable to them as a step toward a much larger goal. This, by and large, has not been done well anywhere except in the large plantation industries, some agricultural industrial research programs, and the large international foundation research programs where people are paid well to be cooperative not only in their research but also in all other efforts.

Certainly, federal or state institutions have had great difficulty in solving this type of problem and most have never seriously tried to solve it. Somehow the "publish or perish" philosophy so typical of research-oriented institutions must be eliminated, especially among those who are working on the long-term interdisciplinary cooperative efforts to solve large and complex problems. A system of substitute rewards, other than those associated with personal publications, must be developed that will attract outstanding scientists and make them enthusiastic participants in such collaborative enterprises.

As biologists, all of us certainly recognize that human nature and innate "human cussedness" is not likely to change very rapidly. This being the case, the only successful approach is to devise an organizational scheme by which all involved are assured their proper rights, powers, privileges, credits, rewards, and other perquisites.

Chapter 10

Possible Organizational Strategies and Patterns

GENERAL CONSIDERATIONS

The organization of plant protection disciplines through teaching, research and extension has been developed over a period of one hundred years or more in the United States. A system has gradually evolved that organizes all of agricultural teaching, research, and extension under the common unifying umbrella of the land-grant university. It is likely that this organization will continue in the United States since it has been most effective.

Until now the various disciplines involved in plant protection have been located in either the colleges of agriculture or the colleges of arts and sciences in the state land-grant universities. Teaching and research has usually been done in the separate departments, the research being coordinated through the dean of the experiment station having statewide responsibilities. Extension specialists also have usually been housed in the separate departments, but their activities have been controlled and coordinated by an extension director also having state-wide responsibilities. The extension specialist in each discipline, also, typically has had state-wide responsibilities and has dealt directly with the county agricultural agent (or farmer) at the field or farm level. The county agricultural agent(farm advisor), then, for the most part has been the "broker" or arbiter for the farmer. This role functioned very well when life and agriculture were somewhat simpler. Now, however, the county agent's tasks have been complicated so much, and there are so many facets to that role, that as a generalist he finds it virtually impossible to cover all the problems posed by the farmer and his family.

Various new techniques are now evolving for getting information to the farmer more quickly. Some states have gone to area personnel, made up of entomologists, plant pathologists, and the like. These people then work in several counties and function as regional rather than state advisors.

Few, if any, governments as yet have hired generalists who theoretically include in their "bag of tricks" all aspects of plant protection. There are many arguments being waged now for and against the development of such an individual who would function as a general advisor for all plant protection at the regional or local level. Certainly, the general advisor has not appeared as yet in any number although in a few states, on specific crops, he is appearing as a field-level consultant, often as a private consultant.

The different plant protection disciplines in the state universities have largely functioned independently and have carried out cooperative programs usually on a voluntary basis wherever it was felt to be necessary. Actually, this voluntary cooperation has been quite effective at the field level in some states. In those states there is much questioning whether a voluntary cooperative system that is working well should be altered.

Where teaching is concerned, the general pattern has also been an independent posture for each of the disciplines. Cooperative teaching has been rare although a few plant protection teaching programs have been developed recently, and others are being developed. These are discussed in Chapter 11.

Where research is concerned, the cooperative ventures have usually been worked out by individual researchers and seldom, until recently, by edict from a dean or administrator. Largely voluntary cooperative breeding programs, particularly in small grains and corn, have been the rule rather than the exception. These voluntary cooperative research programs have been exceedingly effective in some states. Some of these have gradually developed into large regional cooperative research efforts such as the Hard Red Winter Wheat Improvement Association of the Great Plains states. This has for the most part developed voluntarily because of a felt need on the part of the researchers involved. Such organizations have functioned remarkably well and many would question the wisdom of changing the pattern.

NEED FOR NEW ORGANIZATIONAL PATTERNS

However, there does appear to be a need in some cases for the development of a more formal cooperative approach to plant protection at the teaching and research levels in the American land-grant system. This probably can only be accomplished after a realization on the part of the different disciplines that it is truly necessary, and then only through the development of a strong administrative structure to achieve the necessary cooperative programs. It would certainly be a mistake to make all of the research in these disciplines cooperative. Most of it must and should be of an independent nature because of the specific requirements of the different disciplines and the research needs inherent in those disciplines. However, where cooperative practical control programs are needed on given crops and where cooperative teaching programs are needed for the development of more generally trained personnel to serve largely at the field level, a strategy is needed for the land-grant institutions that so far has not developed in any consistent fashion. Dickerson (1974) made the following comment concerning such a strategy:

> The best strategy for developing pest management programs is to get the Dean, or person of similar stature, to say there will be one, appoint a leader and then back him. Anything less will only

> be partially effective and will wither sooner or later. All the subdisciplines must be represented and have something at stake. The program must be structured so that if any one person or discipline drops-out, the momentum can be maintained until that spot is filled. A group of well-intentioned participants will simply flounder if the authority and to some extent funds, are absent.

This is echoed by Arneson (1975) who has had considerable experience working in cooperative research efforts that include all aspects of plant protection.

It would appear unwise to destroy or alter markedly any of the legitimate disciplines now included in plant protection. They represent natural disciplines and have been very effective as they are. On the other hand, a new approach toward cooperative effort needs to be made that will have strong leadership and that will bring the disciplines together in needed, meaningful cooperative efforts. Perhaps the development of subdivisions of plant protection—within colleges of agriculture under plant science divisions would be a meaningful development. The separate departments of entomology, plant pathology, weed science, nematology and the like would then be under the umbrella of plant protection.

It would be a mistake, and a bad one, to make this type of organization completely separate from a college of agriculture or from the influences of plant science (that is, agronomy and horticulture). What *must* be remembered, even though plant protection personnel often seem prone to forget it, is the critical fact that plant protection has a subsidiary or supplemental role to play in agriculture. It must always be maintained as a part, and often a relatively small part, of the total efforts toward growing a healthy crop. The most important tasks are those of crop improvement, production, and management. In a secondary role, usually more important than plant protection, are the economic considerations associated with growing a particular crop successfully in a particular place. All of these considerations usually must be approached by the grower prior to considerations of plant protection. For this reason the organization of plant protection should be under the larger umbrella organization of Plant or Crop Science.

POSSIBLE PATTERNS OF ORGANIZATION AS CO-EQUALS

There is no question that cooperation eventually needs to be the order of the day among the different plant protection disciplines. The patterns undoubtedly will be different in various states, and in different countries. Good (1974a) outlines some of the possibilities for the organization of delivery systems at the field level. In discussing the situation in this country he points out that the type of crop protection and pest management delivery systems chosen will affect the future control of United States agriculture. His list follows:

(1) Private consultants or private consultant firms.

(2) Commercial consultants or commercial consultant firms.

(3) Retail pesticide salesmen and applicators.

(4) Technical representatives of the agricultural chemical industry.

(5) Grower associations or cooperatives operated and controlled by growers with the development of pest management districts or similar needed organizational patterns, but all under the control of a grower group.

(6) Corporate farms or large farms big enough to hire their own crop protection and pest management personnel.

(7) Public or government pest management programs operated by the extension services, or a similar service, and utilizing state, county, city, and possibly even community staffs not only for crop protection and pest management but possibly also for regulatory activities.

He suggests that in the United States probably one or more of the above approaches will be developed, and he thinks that the private sector will be more important in whatever delivery systems are eventually devised. The public sector, he feels, will continue largely in an advisory role. He expects the organizational approaches to vary depending upon the state, the crop, the acreage of the crop, types of pests, and the kinds of control programs that are required. Where small, low-income farmers are involved he thinks most of the crop protection and pest management advice they receive still will come from the public sector through the extension service.

Fitzsimmons (1972), who is reasonably representative of the agricultural chemical industry in the United States, discusses in the following statement the kind of delivery service organizations that can be expected if the agricultural chemical industry is involved:

> Whoever sells such a service would presumably be specially trained, licensed and bonded. Agricultural pest control would then be handled as human and veterinary medicine, the function of the pharmacist and doctor being combined in a single individual, who might or might not be an employee of the pharmaceutical company. Carrying the analogy one step further, we cannot subscribe to a system of socialized agricultural medicine where all pest control responsibilities are given to government employees who might well become legalized regional czars.
>
> The agricultural chemicals industry would prefer to maintain its role as supplier with responsibility for pest control practices remaining with the farmer. However, if the public and the regulatory agencies demand a radical change of systems, industry will consider the assumption of pest management or even crop production contracts, provided that their extra services are paid for.

This statement raises some interesting and serious questions. First, if an agricultural industry assumes the role of advisor and also applicator, can it be expected to be objective about the use of its own materials? Many question the capacity of even well-intentioned agricultural chemical companies and their personnel to be completely objective about the use of chemical materials, especially their own. It would appear that the grower needs an "independent broker" who neither represents the government nor industry. This would seem to point toward the development, particularly in the case of large-scale modern agriculture, of private consulting firms or individuals to function as the objective arbiters concerning questions of alternative control methods.

In the United States it would appear that the role of the extension service would continue to be largely advisory, hopefully continuing to give objective advice, as has usually been the case in the past, and in all probability serving as the "broker" or arbiter for the small or low-income farmer.

In the developing world, which is made up largely of small, low-income farmers, it would appear that the pattern developed in the Republic of China (Taiwan) would be worth serious consideration. In most developing countries the small farmer is essentially ignored now.

By contrast, most of the large plantations—banana, pineapple, sugarcane, citrus, tea, coffee, and the like—are large enough to hire their own plant protection personnel. This pattern for large plantation industries probably will continue in the tropics since it has been quite successful in the past.

It is interesting to note that the Republic of China (Taiwan), one of the tropical countries that has made very rapid advances in recent years and that is characterized by many small, relatively low-income farmers, has devised a very sophisticated advisory and pesticide application service operated completely by the government extension services but controlled through farmers' private organizations; a very interesting combination of cooperative private control utilizing public funds and personnel. The only place private consultants are used in Taiwan is among some of the relatively large, high-income growers. Among these growers a few private consultants and applicators have developed businesses (Chiu 1974).

INTERNATIONAL ORGANIZATIONAL EFFORTS

Several United States universities have developed essentially international plant protection programs that are aimed at cooperative efforts in plant protection, and which have functioned primarily as AID programs for the developing world, either in research, teaching, extension, or a combination of these. The International Plant Protection Center at Oregon State University, developed under the leadership of

W. R. Furtick (formerly Chief, Plant Protection, FAO, Rome, and now Dean of the College of Tropical Agriculture, University of Hawaii), is an example of one of these. Its primary thrust is in weed science.

The University of California at Berkeley has a pest management and environmental protection project. It is the contractor for worldwide surveys to identify priority problems and to develop long-term strategies for providing U.S. assistance in plant protection and control programs in teaching, research and extension. The study teams making the surveys are multidisciplinary and include weed scientists, entomologists, nematologists, plant pathologists, and others. Six study teams have been set up and the recommendations of the survey teams will be designed to help in organizing and catalyzing the activities in pest management and environmental protection in countries that are surveyed. The information developed from the surveys will also help existing programs in the United States and hopefully improve their competence.

Other land-grant universities in this country also have programs in plant protection in various parts of the world. These are usually associated with larger agricultural programs that involve all aspects of raising an important crop or group of crops, certainly a better organizational pattern than considering plant protection as a separate entity.

In the developing countries, which do not have a long history of effort in plant protection, the barriers between the disciplines are not as great, and the barriers to formal cooperative effort in the scientific community usually have not been built up over the years. Since there is no history of large-scale plant protection efforts in most of these countries except on the large plantations, they are actually in the process of developing something new. In the Asian world most countries seem to be moving toward institutes of plant protection that include all aspects of plant protection. In contrast, there is a scarcity of similar effective efforts in the United States where plant protection has a long history and has developed independently in the separate disciplines.

It is interesting to note, too, that representatives from the developing world usually want to see the development of a doctorate in plant protection, somewhat like the Doctor of Veterinary Medicine, to function as a general practitioner or generalist at the field advisory level. By contrast, in this country most leaders of plant protection disciplines are very lukewarm or antagonistic to this idea (Hess 1974). These reactions to the development of plant protection as a separate discipline were very apparent in a meeting of agricultural administrators from Asia and the United States held in 1974 by the Food Institute of the East-West Center in Honolulu.

Both Wharton (1969) and Apple (1972), among others, have stressed the extreme importance of the development of plant protection centers in various countries of the developing world. Wharton (1969) has emphasized the fact that the "green revolution" could well fail because of the inability to control plant diseases and pests. Apple (1972) stresses the high crop losses through pests and plant diseases and the dangerous misuse of pesticides in the developing world.

Fortunately, there is now under development the extensive program in plant protection being sponsored largely by FAO, United Nations Development Program (UNDP) in cooperation and coordination with various countries of Asia. The first such plant protection center was initially planned cooperatively in the Republic of China (Taiwan) and was established in July 1971 under a technical cooperation agreement with the Republic of China (Taiwan) and the UNDP. This was reorganized in July 1972 under the Joint Commission on Rural Reconstruction (JCRR) and is now being operated under the aegis of the Taiwan government alone. Its present director is Dr. Ku-sheng Kung and its purposes are to develop integrated systems of pest control (for plant diseases, insects, rodents, and weeds), to emphasize basic and applied studies on pesticide residues and pollution problems and ways to minimize the use of agricultural pesticide chemicals, and to develop phytopathological and entomological investigations dealing with diseases and pests of major economic crops. At present this organization has five divisions: Pesticide Residue, Pesticide Toxicology, Entomology, Plant Pathology, and Plant Physiology. It now has a staff of 42 including 26 research workers, 7 of whom have doctorates. It also is associated with research, training, and extension programs in cooperation with other government agencies and universities in Taiwan (Kung 1975).

Other plant protection research and training centers similar to this are being sponsored jointly by the FAO (UNDP) cooperating with other countries in Asia. All of these are being developed around the concepts already outlined in the Taiwan program and include research, teaching, and extension components. These are being carried out not only in cooperation with local governments but also with local universities. Thus far plant protection centers have been initiated in Korea, Thailand, India, and Indonesia. They are being planned for the Philippines, Burma, Sri Lanka, Afghanistan, and perhaps elsewhere. This is a most important development and should help each of these countries tremendously in developing unified, cooperative efforts in plant protection (Furtick 1975).

Also, cooperative plant protection efforts, usually on single crops, are being sponsored effectively in developing countries of Asia and elsewhere through large commercial plantation organizations. This is also true of the international foundation sponsor organizations such as the International Rice Research Institute in the Philippines. These usually concentrate on a specific crop or a small number of crops. National agricultural research institutes also are functioning in the Philippines, in India, and elsewhere. Several agricultural institutes that are built around a given commodity or crop are being developed or have been developed recently in Indonesia. In each case these organizations specialize in team research and include all of the agricultural disciplines, including plant protection. These apply the team research approach on a specific crop. The primary emphasis here is, and of course should be, crop improvement, production, and management. Anything that is associated with the improvement of any of these goals for the specific crop is considered to be justifiable research. Individuals in plant protection are part of each of the teams and fulfill their proper role as supporting members of a team concentrating on the improvement of a given crop. This organizational plan has been extremely effective world wide and it probably does not matter whether it is sponsored by a commercial plantation or company, a foundation or a government, as long as outstanding personnel are involved in well-financed team research efforts. The "green revolution" in the developing world is largely an outgrowth of team research of this type working on the improvement of a given crop. Parenthetically, this is also true of the earlier "green revolution" associated with hybrid corn development and small grain improvement in the western world. In the United States much of this early research was done by individuals working voluntarily in large informal cooperative research efforts.

OTHER ORGANIZATIONAL PROBLEMS

One of the principal problems, no matter what organizational patterns evolve, is to convince farmers that integrated or combined pest control techniques are better than single techniques previously used for plant protection. The farmer is typically a skeptic who has to be convinced, and rightfully so. Certainly, any control techniques of an integrated or combination type have to be as effective and cheaper than control measures developed independently by the subdisciplines, otherwise growers cannot be expected to adopt them. Plant protection is inseparable from economic considerations and good management techniques. For this reason alone, it is a mistake to create separate institutions that are not closely coordinated with the total agronomic or horticultural requirements and practices for a given crop or related group of crops. Also, any control procedure is certainly doomed to failure that does not fit into the soil erosion control practices or other soil considerations required in an area. Any good management practices that are economically acceptable or required by the farmer must be considered first. Plant protection programs must be devised to fit into these economic, agronomic, and conservation constraints.

No matter what the organization, there is also the question

of who pays for the control program. If it is a closely controlled society where government programs pay for the advice and even the application of control procedures and materials, the taxpayers inevitably get the burden. In the open societies characterized by capitalism, the farmer, farmer groups, plantation owners, or large industrial or cooperative farms pay and eventually, of course, the consumer absorbs that bill at the grocery counter.

It is extremely important to remember, no matter what organizational scheme is employed, that plant protection has many ramifications and is tremendously complex. For example, in the developed world, systems analysis and, in general, the systems approach to agriculture and agroecosystems is beginning to be emphasized. This means that not only computer science, but various mathematical approaches to systems analysis and simulation modeling need to be emphasized in training and research programs. It also means that much better training in statistics and applied empirical field plot techniques should be emphasized everywhere. It is at the field plot level that the empirical research information is obtained. This is the ground truth data which can then be made available for any sophisticated computer analyses. We should remember, too, that in most of the developing world the empirical information obtained from carefully laid-out field plot research is the only information that will be available in the foreseeable future. Extensive computer analyses, even if applicable, will not be possible soon.

Many disciplines need to cooperate in plant protection. The principal ones are plant pathology, entomology, weed science, nematology, and soil science. But many other disciplines are involved: agronomy, horticulture, certain areas of zoology, rural sociology, agricultural economics, ecology, plant physiology, biochemistry, and others. It is, therefore, a mistake to develop any organization that excludes any of the legitimate disciplines which should and must cooperate, even though only occasionally, in efforts to develop adequate control programs. Any organizational structures that do not face up to these complexities are doomed to failure, as are any which do not accept the fact that plant protection is only part of the larger effort of crop improvement, production, and management.

Usually it is wise and necessary to organize teaching, research, and extension in separate units working under one centralized system of control. The organization of these three separate units and some possible development of these in the field of plant protection will be discussed in the next three chapters.

Chapter 11

Teaching Problems: Organization and Strategies

GENERAL CONSIDERATIONS

The advanced training of personnel for the various specialties in plant protection has become reasonably well-stylized and fulfills the special needs of each discipline. As has been pointed out many times in this discussion, there are many, however, who feel strongly that people with more generalized training are needed at the field level, people whose training cuts across traditional discipline lines. They feel that someone equivalent to the general practitioner in medicine is needed at the field level, a person whose background includes all the various fields of plant protection.

This problem is discussed in the following quotation (Anon. 1972b):

> The availability of specialists is limited by the number of institutions that offer appropriate training programs, which, in turn is affected by the demand for integrated pest management. Qualified individuals are not yet available in many areas of the country (USA). Crop protection specialists require a broad understanding of pests including natural control agents and

other environmental influences; economic thresholds; crops and modern farming practices; and complete and up-to-date knowledge of control measures. Training programs for these individuals require heavy emphasis on a number of disciplines in the physical, biological, and agricultural sciences as well as extensive field experience. Not many will undergo the rigorous training without some assurance that integrated pest management offers a career opportunity. In addition, individuals qualified as crop protection specialists will want some recognition of their training and will need protection from criticism because of the practice of less-qualified individuals. Such recognition of qualifications is also important to a potential crop protection specialist because it may influence his ability to obtain adequate liability insurance coverage.

Hooker (1975) presents the problem in another way:

> We need people who are crop and agriculturally oriented, rather than academically discipline oriented. I believe that people and the environment in which they work are important in the development of cooperative and coordinated approaches to pest control problems. For this, I believe we need to give graduate students broad training including entomology, plant pathology, and genetics. They should also have field experience working with crops. Too many of our plant pathology staff and students are laboratory oriented. The plant breeders, agronomists, plant pathologists and entomologists are often in the same building and work nearby in the same greenhouses and field plots. They should also be in constant contact with the farmers and industry if they are to make cooperation work. Administrations can discourage this cooperation by their actions much more effectively than they can encourage it.

Some think it is impossible for one man to absorb all the information in these different disciplines and be effective in the whole area. The arguments continue. Actually, we do not know how effective individuals will be in attempting to cover all field aspects of plant protection. One thing is certain, there would be few, if any, individuals who could do this without an enormous amount of help from specialists in the disciplines. It is clear that a relationship will need to be developed in which the generalist working at the field level has quick access to help from specialists when needed. This relationship would be much like the relationship between the general practitioner in human medicine and the medical specialists.

Plants at times are adversely affected simultaneously by insects, diseases, weeds, nematodes, vertabrate pests such as rats, and several abiotic maladies. Diagnosis in such cases becomes exceptionally complex and consequently the plant protection man working in the field certainly needs to be broadly trained (Heagle 1973). The field man being described in the previous statements must be a very well-trained person, one who will never exist in large numbers unless means are developed by which he is given proper legal protection, adequate prestige, and high pay. At the moment the prestige level for such a worker is so low that it is very difficult to get people with outstanding ability, ability enough to absorb and use all the necessary information, to go into such a profession in the first place. This problem as it relates to foreign students in the United States is discussed in the following quotation (Anon. 1968).

> The problem of degree of orientation toward either basic or applied effort is reflected in the diversity of training programs from which our future manpower must come. The large graduate training centers are under constant pressure for more basic orientation, pressure that stems indirectly from competition for research support. The more gifted students are well aware of the growing shortage of scientific manpower and have been increasingly aligning themselves with fundamental research because of interest, professional status and security. This situation poses a very difficult problem for those graduate training centers trying to maintain programs that will accommodate applied or descriptive graduate theses, particularly for the foreign student whose employment opportunities at home do not yet require the most highly sophisticated training.

As implied in the foregoing statement, the power, prestige, pay, status and most of the desirable personal goals for the scientists in plant protection disciplines are stacked very strongly in favor of the so-called fundamental or basic research effort. Unfortunately, people who make up this group also often attempt to dominate the direction of applied research though they, for the most part, do not understand the problems nearly as well as those working at the field level. Although much lip service is paid to supporting those who work at the field level in applied research and extension, actually the prestige of the disciplines is centered in those who control the most sophisticated equipment and do what is presumed to be the most sophisticated, fundamental research, largely in laboratories. As long as this situation lasts, it will be extremely difficult to develop adequate numbers of outstanding general practitioners to function largely at the grower and/or applied research level.

Apple (1974) and several others have stressed the fact that the traditional undergraduate programs in entomology, nematology, weed science, plant pathology, and so forth have rarely stressed the integrative or ecological approach in plant protection. Although a number of land-grant universities have developed plant protection "types" of curricula both at the undergraduate and at the graduate level over the past few years, these curricula generally are interdepartmental, involving weed science, agronomy, entomology, plant pathology, and sometimes other departments, and they typically use courses already in the catalogue. Only rarely have special integrated crop protection courses been developed. Apple reports that there is a consensus among those who advocate integrated approaches to pest management and cooperative approaches to all plant protection that new courses will be needed that approach the disciplines in a coordinated fashion. These courses, he feels, are particularly needed for personnel being trained for the extension service, private consulting activities, and the agricultural chemical industry. He feels that programs need to integrate the commonalities of the various disciplines and give the fundamentals of the crop protection-pest management disciplines as well as economics and mathematical modeling. This program as described by Apple is a very intensive one which probably would have to be conducted at the master's degree level. One wonders how many people with adequate ability can be persuaded to enter this type of training when the pay, prestige and job prospects are still in doubt.

In recent years the Resident Instruction Committee on

Policy (RICOP), Division of Agriculture, National Association of State Colleges and Land Grant Universities, has been studying the possibility of the development of "Pest Management for Crop Protection." They have developed curricula for the bachelor of science degree in which very broad coverage of the various disciplines is given and have recommended a six months practical internship. They have also recommended the development of master of science programs characterized by advanced training in all disciplines represented and the use of broad concepts coming from all disciplines. Each discipline would be represented on the graduate committee.

They have recommended that a doctoral program be developed along traditional lines as part of existing programs, but that it include field-oriented research in which a systems approach is utilized, and that all areas of crop protection and pest management be represented on the graduate committee. They have also encouraged the development of short courses and seminars for supplemental training and a postgraduate one-year course to retrain people with advanced degrees for new pest management jobs. In these programs the systems approach to problem-solving would be emphasized (Anon. 1972d, 1974).

Osmun (1972) disagrees completely with the RICOP group recommendations concerning the bachelor of science degree. He feels strongly that the type of training possible at the bachelor's degree level will mean that individuals are inadequately trained in all fields. This type of training, he says,

> spreads a man so thin he isn't competing in any phase of pest management. A single person cannot do everything; pest management will succeed only when a number of people, each with his own capabilities, are brought together as working units to implement a program.

Osmun feels the important training procedure for crop protection and pest management is to retrain many of today's extension entomologists, plant pathologists, and so forth. This retraining should include ecological principles, systems analysis, manipulations of life systems, and other pertinent subjects. He feels that a hierarchy of people are needed in plant protection at the field level and thinks these people can be trained with much less effort than is necessary for training those at the bachelor of science level. His hierarchy would be as follows:

1) Laborer or scout: This person functions in the field on a specific crop and could be given on-the-job training for the necessary diseases, pests, and so forth, at the beginning of the crop season. He feels that intelligent high school graduates, students going to college taking science or agricultural courses as majors, could be trained adequately for this work and, actually, they have been trained adequately in several scout programs now in existence.

2) Applicator: This person, Osmun feels, needs technical and operational training in the use of materials, equipment, and the restrictions concerning the use of particular chemical materials. He feels these persons can be trained adequately by attending workshops or short courses given by extension personnel working out of land-grant institutions.

3) Supervisors or Special Application Personnel: He feels these people working as supervisors of applicators would need to have formal training, perhaps one or two years of training in extension courses and particularly short courses. They would need not only technical competence in handling chemical materials and equipment but also would need to be certified. They would need to have a reasonable amount of leadership ability for organizing applicator units and controlling laborers or scouts in the field.

4) Operational and advisory personnel: These people should have college degrees in an appropriate specialty and also certification. The master of science degree is probably the minimum level of training required. These would be extension workers, industrial personnel, private consultants, and so forth. Osmun does not think the bachelor of science degree program would be sufficient training for these operational and advisory personnel.

In December 1972 a workshop on "Pest Management

Curriculum Development and Training Needs" was held at the Food Institute of the East-West Center in Honolulu. At that time tentative curricula for the bachelor's degree and possible master's and doctoral programs in crop protection and pest management were developed, as well as non-degree training programs in pest management. These tentative curricula, as well as the discussions concerning them, have been made available to land-grant universities in the United States and to equivalent institutions throughout Asia. In this and other conferences and seminars held concerning pest management at the Food Institute, it has become clear that Asians and generally those from the developing world would like very much to have a doctorate in plant protection, somewhat—as mentioned earlier—like a doctor of veterinary medicine degree. They would prefer that it be developed in several land-grant institutions in the United States or similar institutions elsewhere in the developed world and be made available for personnel from the developing countries of Asia. It is equally clear that those representing the land-grant institutions from the United States typically do not favor this idea. No such degree programs have been planned or initiated thus far (Hess 1974).

TRAINING PROBLEMS IN THE DEVELOPING WORLD

When we go to the developing world and examine its education and training problems, it is clear that some of these are different from those in the developed countries. Paddock (1967), a man with a lifetime of experience working in the tropics in plant pathology, gives the following analysis of the needs and problems:

> What we need is a quick way of training a corps of plant pathologists, entomologists, agronomists and so forth within the developing countries. The headlong rush of the food crisis calls for agricultural competence within a decade, not within the next century. About the same method of training plant pathologists and other agriculturalists is now used in Nicaragua and Nigeria as in England and the United States, 9–12 years of grade and high school, followed by 4–6 years of university and then

followed by graduate training. It is a good system if there is time, as there is in England and the United States. There isn't time in Nicaragua and Nigeria.

Do not look for an agricultural revolution in the hungry world, there is no one to lead it. The future looks no better than the present. In nearly all Latin American countries, for example, the percentage of university students studying agriculture has decreased during the last decade. For instance, Mexico has declined from 3 to 1%, Panama from 4 to 2%, Dominican Republic from 2 to 1%. Of 105,000 Latin American students enrolled in the United States during the decade 1955–1965, only 5% studied agriculture. Some way must be found to make available to the developing world the skills of the plant pathologists and workers in related sciences without men having to pass through the usual educational system. How can it be done? No one now knows the answer, but has the question ever been seriously faced by a competent group of men. Possibilities that come to mind are:

1) teaching via television (perhaps even by satellite transmission) to compensate for the teacher shortage;
2) mass use of rural (transistor) radio to substitute for the nonexistent or immature agriculture extension agent.

Paddock describes a mammoth problem, one which no government has really faced up to completely. It also shows clearly again the extremely low prestige of agriculture in the developing world. Unless many outstanding minds can be enticed into scientific agriculture, the situation will go from what is already very bad to what will undoubtedly be much worse.

A suggestion made by some is that an international center for practical agricultural training be developed, including all crop protection and pest management specialists, perhaps with its headquarters and activities coordinated by FAO, Rome. Utilizing satellite training possibilities and rural transistor radio training, as suggested by Paddock, this global network could be designed not only for quick training of personnel but also for quick response to field problems of great practical significance in all countries of the developing world. This would be a teaching, applied research and extension information center aimed at the global crisis in agriculture. Response to the global crisis should certainly include the rapid education of needed personnel for work on important problems in plant protection on the important crop food plants now in short supply. Using known techniques and a worldwide network, it would be possible to obtain the services on a routine basis of the world's most outstanding applied research field practitioners. They could teach courses beamed from FAO headquarters, Rome, or elsewhere, and help in responding to crisis field problems reported from all over the world.

Buddenhagan (1975) also emphasizes the differences in the requirements for training plant protection personnel in the developing world as compared to the developed countries.

> There are great differences in the requirements for generating agricultural improvement in the tropical (Asian) countries and in the requirements for such in the United States and Japan. These differences apply to
>
> 1) kinds of research needed,
> 2) kinds of extension needed,
> 3) kinds of education needed.
>
> For those who are struggling with problems in tropical countries there is a fundamental pitfall in the education of their sons and daughters in "advanced" temperate countries. For those in "advanced" temperate countries, there is a fundamental pitfall in their attempt to help educate people from tropical countries.

So often, as Buddenhagan implies, students from the developing world are given such sophisticated training in the developed countries that when they get back to their home countries, they cannot actually use their training in a creative way. In reality, what they need for the problems of their countries is excellent training in adaptive applied research at the field level. Often these individuals know nothing of field research when they return to their countries and even less about extension methods and how to apply them. What is often worse, they even spurn or are scornful of such work. In passing, it might be added that it is often difficult in land-grant institutions of the United States to get support for a research project in which a student from a developing country is utilized on an applied research project. It is much easier to place a student in an ongoing basic research project that is utilizing sophisticated laboratory techniques and that is usually not field oriented.

Certainly, the ideas presented earlier by Osmun (1972) can and should be developed as much as possible in the developing world. He suggests developing conferences, workshops, short courses, and symposia on plant protection for the retraining or supplemental training of extension and other personnel who have at least some background in agriculture. These can quickly begin to function on at least one important crop as lower level specialists working with specific plant diseases, insects and other pests. This type of effort is becoming more common throughout the world in educational and research institutions, but many more institutions should be attempting to develop these retraining programs in plant protection disciplines.

A Conference on Integrated Pest Control was held by the International Rice Research Institute May 9–12, 1972, and short courses on the same subject were taught at the College of Agriculture, University of the Philippines, Los Banos, in 1973. The Food Institute of the East-West Center in Honolulu has as one of its central thrusts in pest management the development of conferences, workshops, short courses, and so forth. During the past few years they have had several workshops on pest management concepts, objectives, and problems, and there will soon be several other workshops and short courses aimed at the middle-level worker in this field, particularly for the developing countries of Asia. These programs will be continuing over the next few years, and many more are needed in all countries of the developing world.

EDUCATIONAL NEEDS OF THE GENERAL PRACTITIONER

What are the actual types of knowledge needed by the person working as a broad generalist in plant protection? The question has been asked by many people and there is no general agreement at this time. The emphases that have been variously presented, however, include the following. It is extremely important for a person working in this field to have a good background in horticulture and agronomy. In short, he must know how to identify a normal plant and especially how to grow it. He must know the desirable culture for a given crop or group of crops and the reasons for the soil types and environments needed, the rotational patterns used, time of planting, and so on. He must be conversant with plant nutrition and must be able to recognize problems that are essentially abiotic, environmental and nutritional problems and many others that affect plants and cause severe maladies but which are not caused by living organisms. Unfortunately, many people now working in the pest management field are woefully ignorant about such matters and make ridiculous mistakes because of their ignorance.

Certainly, the basic information in plant pathology, entomology, weed science, nematology, and the like are necessary. Basic courses in each of these fields will be essential. In addition, because of the extreme importance of genetics and plant-breeding in the future of plant protection, not only a basic understanding of genetics will be required but also of statistics and the best applied field plot techniques. Because of the impact of agriculture on the environment, a basic knowledge of ecology is important. Anyone who utilizes agricultural chemicals needs to have a basic background in chemistry and knowledge concerning the problems of residues and tolerances as well as basic toxicology. The utilization of agricultural chemicals becomes particularly complex when several chemicals are used simultaneously. The whole idea of compatibility—the combination of sprays and other materials, and the effect of these combinations, not only on plants, diseases, and pests, but on the environment as well—is an extremely complex area of knowledge that eventually must be understood and become a part of a training syllabus.

A fundamental knowledge of botany, zoology, and physics is also needed, and a basic knowledge of rural sociology and agricultural economics. An understanding of computer science and systems analysis as well as simulation modeling techniques is needed for those going into the more sophisticated applications and implications of field plot techniques and the environmental effects of various pesticides and pollutants. Finally, for those who need to understand the biochemical events occurring in the plant, a basic knowledge of plant physiology, biochemistry and soils is essential.

It is clear that this is a most sophisticated and exacting package of knowledge. Only those few students who are talented and have exceptionally strong motivation will continue to struggle through all of these courses and requirements. This again raises the question of attracting outstanding students to a discipline that has little prestige and not too many immediate job possibilities. In fact, until students can be convinced that a generalized discipline in plant protection offers bright prospects, we can expect very few to enter the field.

TEACHING PROGRAMS IN THE UNITED STATES

Several teaching programs at the bachelor's and master's level have been developed in plant protection in the United States at various educational institutions. The variety of names for these programs indicates the essential confusion and differences of opinion that exist. Only programs for which information has been received will be discussed.

Two-year program. The only institution in the United States known to have started a two-year pest control program is North Carolina State University. This program was initiated in 1974 and the graduate of the two-year program receives an associate degree in applied science and majors in the agricultural pest control option. This training includes the newest biological, cultural, mechanical, and chemical methods for controlling agricultural pests.

Bachelor's degree programs. Bachelor's degree programs have been developed in several institutions. These vary greatly in organization, goals and breadth. Couch (1973) under the auspices of the American Phytopathological

Society has developed a report on the present status of the "crop protection and pest management" teaching programs at the bachelor of science level in the land-grant university system of the United States. Fifty-five universities were surveyed. Twenty-two reported that they were offering or were going to offer an undergraduate program in plant protection. Another 12 reported plans for initiating such a program by 1976. The oldest program was initiated in 1959. Most are recent in origin, only 6 going back prior to 1970.

As indicated above, the most commonly used (22) name for the curriculum is "Plant Protection." Among the remaining 12, either "Pest Management" or "Crop Protection," or a combination of these appear most frequently. Usually these are taught as interdepartmental majors or options within majors under the direction of an interdepartmental committee. A few are controlled by a specific department such as Plant Pathology or Entomology. There are no organized departments or divisions of Plant Protection or Pest Management as yet.

Specific examples of several programs are discussed below.

At the University of Arkansas, a Plant Protection major at the bachelor's level has been initiated which has a very broad base of courses and essentially includes in the curriculum all areas—both biotic and abiotic—of plant protection.

At the University of California, Berkeley, a bachelor of science course in Pest Management was initiated in the fall of 1972. This is a cooperative effort among the departments of Entomology, Plant Pathology, and Parasitology, and includes a pest management one-summer field course of practical training. Its major weakness appears to be in the area of weed science (Couch 1973).

North Carolina State University initiated a bachelor of science degree in Pest Management in 1974. The graduates have a field-oriented approach to agricultural pest management and the emphasis is on problem diagnosis and control methods. This is a joint inter-departmental effort between the departments of Crop Science, Entomology, Horticulture, and Plant Pathology. A new cooperative course has been developed entitled "Pest Management" which attempts to bring together a body of useful knowledge from all of the disciplines. On-the-job pest management training is also stressed and several sessions of practical field training (Couch 1973).

In 1973 the University of Idaho initiated a bachelor of science degree in Plant Protection. The curriculum has a strong background in biology, chemistry, business, and economics. The major fields may be in either weed science, plant pathology, or entomology, and the program represents a cooperative effort between these three departments (Erickson 1974).

Virginia Polytechnic Institute and State University have a broad bachelor of science degree in Plant Protection which has been developed cooperatively by the Departments of Agronomy, Horticulture and Plant Pathology and Physiology.

Kansas State University has just initiated a bachelor of science degree in Crop Protection. The major options may be either pest management, plant pathology, entomology, agronomy, or horticulture. Several new crop protection and pest management courses have been developed totaling approximately eight hours of work (Dickerson 1974).

In the fall of 1973, Oklahoma State University developed a bachelor of science degree program in which a plant protection option or major was included in several departments. The cooperating departments are Agronomy, Botany, Plant Pathology, and Entomology. Students may enroll in any of the departments and take a series of courses required of all students. The two new courses that have been included in this plant protection option are economic entomology, and pesticides and the environment (Santelman 1974). It is interesting to note, in passing, that Sturgeon (1974a) at Oklahoma State University has recently put mobile plant heath diagnostic laboratories into operation, and he has developed what seems to be an excellent practical means of training interns and graduate students in plant health matters at the field and applied laboratory level.

For some years, Oregon State University has had the International Plant Protection Center (IPPC). Its present director is Stanley F. Miller, an agricultural economist. This center is a cooperative effort between Oregon State University and the United States Agency for International Development. It concentrates primarily on research in weed control systems in developing countries, and it also emphasizes the socioeconomic implications of weed control technology on employment and income. A bachelor of science degree is possible in this program. Also at Oregon State University, they are in the process of developing an interdisciplinary course in Pest Management which will be team-taught and which will include weed science, nematology, entomology, and plant pathology. It also will be a bachelor of science curriculum (Miller 1974; Shay 1974).

In Minnesota a bachelor of science program in Plant Health Technology was started in 1973 in the Plant Pathology Department. This is not an interdepartmental program. It is designed mainly to teach the methods of protecting plants from damage by parasitic or microscopic organisms, by environmental contaminants such as air pollutants, industrial waste, and pesticides. It also includes improved management and breeding techniques, biological and chemical control methods that are consistent with the objectives of environmental protection, and a summer field internship. The individuals are trained primarily for industry and government agencies, seed production companies, crop improvement associations, nurseries, forest products companies, chemical fungicide manufacturers, food processors, regulatory agencies, and state and federal government agencies concerned with plant diseases (Wood 1974). This is not a pest management program. The emphasis is preventive and is on the health of the plants. Thirty students enrolled in 1975 and the first graduates have been able to find related employment (Anon. 1975h).

At the University of Hawaii no formal combined training program has been started although one, "Pest Management for Plant Protection," is to be initiated soon. However, for some years Gilbert (1974) has been teaching that many pest problems are too severe or complex to be solved without the application of several methods to the same problem. Twelve general methods of pest and disease control are described in detail to the students in horticulture, and examples are given of successful combinations of control methods. The general methods of control that are considered are (1) crop rotation, (2) resistant plant varieties, (3) biological control, (4) plant quarantine, (5) soil fumigation, (6) thorough plowing and dry fallow, (7) seed treatments, (8) crop abandonment, (9) mechanical barriers, (10) poison bait, (11) trap and catch crops, and (12) spraying and dusting pesticides.

At the University of Wisconsin a new program has been made available at the bachelor's level in the Plant Pathology department. This is not an interdepartmental effort and is the first undergraduate major in plant pathology ever offered at the University of Wisconsin. It is offered under the agricultural production and technology option. It is a broadly conceived undergraduate biological program in plant pathology but also emphasizes agronomy, entomology, horticulture, and soils, and it is designed as a terminal degree. Kellman (1974) is not sure about the implications of this degree and type of training, nor does he feel certain about the job possibilities or the effect it will have on plant pathology as a profession

Master's degree programs. At least four schools have initiated master's degree programs in Crop Protection and Pest Management. Mississippi State has developed a master of arts degree specializing Pest Management. This is a joint degree between the departments of entomology, plant pathology, and weed science and is directed by a committee. It includes 10 hours of plant pathology, 10 hours of weed science, 10 hours of entomology. It is an 18-month program; instead of a thesis, it requires one credit-hour of practical summer experience working with the Mississippi Extension Service. Woodrow Hare, the chairman of the Plant Pathology Department, indicates that the first graduates of this program are getting good jobs at satisfactory salaries. At Mississippi State University, they feel that this is a simple and effective way to broaden the training base, and that graduates will be able to obtain good positions (Blasingame 1974; Hare 1974).

At the University of Illinois, a master of science degree is available in Pest Management through the department of Entomology. This is not an interdepartmental program and does not include all disciplines. It does include the integrated pest management concepts and includes new courses in ecology, economics, and sociology but little in other plant protection disciplines. It includes a thesis and field training of at least one semester working in the Illinois Natural History Survey. The program was initiated in 1974 (Larson 1974).

At Virginia Polytechnic Institute and State University the Department of Plant Pathology and Physiology has developed a non-thesis master of science degree in Plant Protection. It is a single academic year program designed to develop basic knowledge and ability in plant/pest control while maintaining some flexibility for students based upon their interests and background. It includes practical training in field, greenhouse and laboratory. It does not include weed science work. It places heavy emphasis on the area of plant pathology and includes only one entomology course (Couch 1974).

Several other courses are available in pest management in other universities. These have been developed, primarily, by entomology departments and emphasize entomological training.

Doctoral programs. As far as known, only one doctoral program is contemplated in the United States, that is at Kansas State University. This program, if developed, will be in Crop Protection. It will be a total plant protection program and will include all areas. It will be an interdepartmental program controlled by a committee. Brigham Young University and Kansas State University are now in the process of discussing this possibility as a cooperative effort (Hess 1974).

As indicated earlier, thus far no university in the United States or elsewhere, seems to be interested in developing a special doctoral degree for a general practitioner, such as Doctor of Plant Health, Doctor of Pest Management, Doctor of Plant or Crop Protection.

TRAINING PROGRAMS WORLDWIDE

In other parts of the world, there have been efforts toward the development of interdisciplinary plant protection teaching programs. In Canada, at Simon Frazer University, British Columbia, a program was started in the fall of 1973 which awards a Master of Pest Management degree. This is an ecologically oriented management program that includes heavy emphasis on biological control and very heavy emphasis on entomology. Much lighter emphases are placed on nematodes, plant pathogens, and weeds. A little effort is directed at the problems associated with harmful vertebrates. It includes two semesters of course work and one summer of practical instruction and field work.

In 1972 the International Agricultural Center at Wageningen, The Netherlands, gave a first international course in plant protection which lasted about four months. This was a good all-inclusive type of course which included essentially all areas of crop protection and pest management, but the treatment, of necessity, was relatively light, because of the shortness of the time. Another session of this postgraduate training program was held July 29–November 20, 1975. This type of program will be continued periodically at Wageningen (Anon. 1975i).

The University of Bath in England has developed a four-year honors degree in Applied Biology equivalent to the bachelor of science or, perhaps, the master of science degree in the United States. This course produces a Crop Protection Advisor and includes training programs with selected commercial firms and public-financed agricultural advisory centers as well as research institutions. It includes very broad training and a background in basic and applied agriculture as well as plant protection.

Plant protection training programs are being initiated as part of the program of the various Plant Protection Centers being developed by FAO (UNDP), United Nations, Rome, in various cooperating countries in Asia. These centers will include all aspects of plant protection. A Plant Protection department with a major was recently started at the College of Agriculture, Baghdad University, Iraq. The name "Plant Protection" typically appears in the curricula from Russia and China and most areas of the world outside the United States and Canada. Where used, the term "Plant Protection" typically includes all aspects of crop protection and pest management except the animal and human aspects. Plant Protection departments also have been developed recently in universities in several European countries.

In the Republic of China (Taiwan) there is degree training at the bachelor's level in pest management at both National Taiwan University and National Chung Hsing University. Approximately 140 students graduate each year from both of the schools. A master of science degree is available at National Taiwan University and is taught cooperatively by the departments of plant pathology and entomology. Some sixty students have completed this degree since it was first initiated in 1947 (Chiu and Yen 1972).

Chapter 12

Research Problems: Needs, Organization, Strategies

GENERAL CONSIDERATIONS

A large amount of independent research has been done in the various disciplines associated with plant protection. Some of this research, as we have seen, has been done cooperatively. However, a vast body of unrelated, independent research has accumulated that thus far has not been utilized or organized in any practical way. As we have emphasized again and again, there is a great need for meaningful synthesis of this body of research to make it available for practical use. Campbell (1972) discusses this problem:

> In my opinion much of the research ostensibly directed toward understanding mechanisms by or through which specific parts of the system operate has in fact preceded with no clearly defined direction whatever. Too much of the research effort in the past seems to have been based on the premise that "someday someone will put it all together."
>
> Unfortunately, this premise has not proven to be true. Our best hope, I think is that in the future we will define both our goals and our operating framework clearly enough in advance

that discoveries regarding the mechanisms that drive individual parts of the system can be placed in the context of that system immediately rather than hoping for that great "someday" when everything will magically fall into place.

Campbell discusses research related to systems analysis and the utilization of computers in developing these analyses, but it applies equally well to a great quantity of accumulated research knowledge that has not been coordinated in any way with other related knowledge. It is usually difficult to obtain research support for the development of monographs or symposia, but without question one of the most significant types of work that can be done at present is to make major efforts to bring together in meaningful units the many bits and pieces of research on the same subject. This should be part of an organized drive to find out what all of these efforts mean in relationship to possible solutions to practical complex problems.

One of the most difficult problems in plant protection is to achieve a meaningful and realistic balance between applied and basic research (Anon. 1968). It certainly is important that investment in basic research be increased to provide new basic knowledge. It is also equally important to invest in applied research so that the basic knowledge can be utilized as quickly and as practically as possible. It has been extremely difficult to achieve a balanced approach although, in reality, basic and applied research are like two sides of the same coin. Granting agencies should consider them together rather than separately. Unfortunately, granting agencies are usually dominated by people who are primarily interested in basic research. It has been exceedingly difficult to convince these people to maintain a balanced posture in relationship to applied research, in particular to research that is related to the practical problems of agriculture.

INTERDISCIPLINARY RESEARCH NEEDED

In the future it is especially important to emphasize interdisciplinary research and interdisciplinary cooperation between the different plant protection disciplines (Anon.

1968). It has been noted that specialization is increasing as it undoubtedly should. But it is extremely important for personnel in the various plant protection disciplines to make every effort to join in cooperative research with not only other disciplines in plant protection but also with plant physiologists, geneticists, biochemists, soil and crop scientists, and others. Actually, it has been exceedingly difficult to develop meaningful formal interdisciplinary research programs except in a few cases where such research has been set up in an interdisciplinary fashion at the outset. The large tropical plantations and the large international foundation-supported research programs in agriculture would be good examples.

Another very difficult administrative and personnel situation has been experienced by many who have attempted to develop new approaches to research in plant protection. This problem is emphasized in the following quotation (Anon. 1966):

> Though we as scientists may decide what should be done and how, it is the people at higher levels in our organization, the people who control the money who decide whether we can do what is needed. The important point is that these people must be persuaded or influenced or educated to recognize that requirements for research in integrated control (interdisciplinary research) can differ markedly from those in more limited areas of control research.

In particular, most interdisciplinary research aimed at coordinated or integrated control programs is of the type that not only must be done by quite a few people but must be done over a considerable period of time before meaningful data can be accumulated. Much effort is required to organize personnel, do the research, and collect and sort the data. Such interdisciplinary efforts are not likely to produce publishable results for some years and administrators tend to feel that the absence of published material is evidence that workers somehow have not been busy. The "publish or perish" point of view is not conducive to the development of long-term interdisciplinary research efforts. There is also a tendency for people involved in these interdisciplinary programs to be promoted more slowly than their peers in other areas, to be limited in supplies or facilities as the work progresses without publication, to be held in less esteem with minimal prestige by their peers and administrators. In short, the whole attitude toward interdisciplinary research has been unfavorable for its development, particularly in the large land-grant institutions in the United States. Typically it has been the well-supported, isolated, and usually competent "prima donna" in basic research who has been able to publish faster and who has gotten the promotions and the prestige. Until this picture is altered it is not likely that we will see large cooperative interdisciplinary applied or adaptive research programs developing on specific crops, particularly in programs that are also tied carefully into considerations of ecology, economics, toxicology, environmental pollution, and so forth.

Many requirements for these large research programs are unique. The motivational approach, the personal rewards, the possibility for publications, for prestige and the like are different, and the organization and administration are often complex. Administrators who are interested in solving some of the complex field problems by interdisciplinary research need to take a hard look at the way in which they reward and support people who go into such programs. Until this problem is faced honestly and creatively by administrators, it will be very difficult, if not impossible, to develop long-term meaningful programs (Anon. 1966).

The following quotation (Anon. 1966) highlights some of the very important problems that must be faced by administrators in assessing results of long-term interdisciplinary research.

> The absence of publications must not be regarded as evidence that the workers are not working, and must not retard their promotion or limit the supplies and facilities needed in the work. These two needs, for large numbers of assistants and for support during the initial apparently unproductive years must be provided the scientists by the administrators if the work is to be done properly and at an adequate rate of speed. We will have

> to educate the administrators appropriately, which may not always be easy.
>
> The development of practical integrated control programs will create a third need that the administrators will have to face, the need for an increasing number of experts to apply them or to direct and supervise their application because many integrated control programs are likely to be too complex and sophisticated for untrained people to apply effectively on their own initiative. The administrators who control the supply and use of money can be influenced by examples of past accomplishments and by examples of barriers to what could be accomplishments.

Certainly the "publish or perish" philosophy as it is related to promotion, pay, and prestige is at the heart of this problem. This philosophy was not always present. In fact, it appears to have been encouraged by some of the pioneers in the United States in plant protection disciplines. One of the early workers, L.R. Jones, emphasized the importance of publishing results quickly (Jones 1919). As pointed out by Bennett (1973), most of the earlier plant pathologists were not inclined to publish results quickly and there was far less economic or prestige pressure for publication. The early research scientists were essentially free to pursue their investigations for extended periods of time before presenting the results of their research in published form (Bennett 1973). There certainly must be some logical position between publishing quickly, almost feverishly, and publishing at a reasonable time at the end of a meaningful large unit of research. Certainly, in the large interdisciplinary research programs it is not going to be possible to publish often, and when the results are published many of the people who have been extremely important in the research will appear to have been left out altogether. Some mechanism of rewarding such research personnel is exceedingly important and must be developed by administrators. Otherwise it will not be possible to convince outstanding people to enter such interdisciplinary research programs. The pay, the prestige, and other perquisites *must* be equivalent to those enjoyed by the individuals working in basic research.

One of the most important needs in interdisciplinary plant protection research is a searching examination of the economics of all plant protection procedures. This was emphasized by Van der Plank (1963), quoted below, and Chant (1966) concurred:

> The first and most important research need in the synthesis of a pest management program is not primarily biological at all. It is economic. A clear picture of the complex economics associated with crop production is vital. First, one must know the general economic picture and then determine what might be called the economic degrees of freedom. In other words we must determine the margin of profit on which the agriculturalist or forester is operating so that we can calculate how much he can stand to lose through the depredations of pests. Secondly, and against this background, one must determine how much can be afforded for protection against this level of loss. This knowledge defines our problems as entomologists and sets the limits on the costs and value of the management programs we can develop.
>
> There have been few analyses of the economics of crop production relative to pest managment problems, and too frequently they have embodied only a descriptive account of the funds used by an agriculturalist in production against the background of his total operation. Principles have rarely been developed and limits clearly defined. Consequently, we have not infrequently been treated to the absurd situation where more is spent to control a pest than the value of the commodity the pest could destroy, or even worse, where a helpful insect is destroyed at considerable cost. An example of the latter is an insect that attacks fruit early in the year and actually benefits the grower by advantageously thinning his crop. It is ridiculous to spend money to apply a pesticide to destroy this insect and then immediately turn around to hire labor to thin the crop.

Few people working in plant protection are trained in economics and much more interdisciplinary research needs to be done with economists. As Van der Plank (1963) reminds us "overriding all is economics." In particular, the concept of the economic threshold needs to be emphasized much more in plant protection than in the past. Apple (1974) emphasized the importance of research on economic thresholds:

Pest management principles cannot be applied without economic thresholds to guide in the development of optional management tactics. The development of economic thresholds requires biomathematical and economic expertise on the pest management team. The threshold concept has been applied widely to insects and to some extent to nematodes but has not yet been applied widely to disease organisms.

Chant (1966) noted this point also:

It is exceedingly difficult to determine the economic thresholds of most pests on most crops. To do so requires an ability to predict the probable consequences of continued increases in populations if controls are not exerted, in relation to the subtleties of injury levels, that is usually beyond our competence at present. The gathering of data for predictive purposes is one of our most pressing research needs at this time. The major difficulty in determining the factors in an ecosystem that regulate pest populations, or have the potential to do so, is that of gathering the complex data required and of analyzing them to extract the information required. One particularly useful tool is now available for this (in entomology), the life table. It is difficult to believe that satisfactory control systems can be developed without understanding the dynamics of the pests of major concern.

When one considers the difficulty and complexity in plant pathology of developing disease forecasting systems and of determining with accuracy the epidemiology of each plant disease and the bionomics of its causal organism, one can begin to appreciate the amount of research time that is going to be necessary to gather adequate data for predictive purposes in the general area of economic thresholds of plant disease organisms. Here, too, the kind of research that is needed is a complex interdisciplinary mix requiring long-term commitments of personnel with little likelihood of early publishable results.

Another area where major research efforts are needed very badly is ecosystems analysis. The systems approach to research is relatively new and has been made possible through computer technology. This also is, of necessity, interdisciplinary research and is likely to be the type which not only takes much time and many people but also from which only a few, hopefully outstanding, publications can be expected. As Van der Plank (1972) points out, it is extremely difficult research:

> Our knowledge of the interactions considered qualitatively embraces much of our knowledge of plant pathology, entomology, and related sciences. It is when one starts probing quantitatively that one becomes intensely aware of ignorance. In plant pathology, particularly, knowledge of even the most elementary relations is vague and inadequate. For example, possibly the simplest relation we ought to know in the whole ecosystem of plant disease is the relation between inoculum and disease. If one hundred spores having fallen on a leaf start *n* lesions how many lesions would two hundred spores start? Could any question be simpler or its answer more fundamental to quantitative analysis of disease systems? But we cannot answer it.

For a variety of reasons, ecosystems analysis has not been popular thus far in departments of plant pathology and certainly not much more popular in other plant protection disciplines. Van der Plank (1972) emphasizes this point:

> Ecosystems analysis (embracing the epidemiology of plant pathogens and population dynamics of insects) has been the "Cinderella" in many departments of plant pathology and entomology. An imbalance between chemistry and mathematics as basic sciences in these departments may have helped to make it so. Basic research is needed. For too long in some institutions basic research has been interpreted as research not inspired by farmers' needs and without foreseeable applications to agriculture. All this I believe to be changing. Experience shows it usually takes more intellectual prowess to solve a problem of immediate practical importance than one judged mainly as a way to scientific publications.

Along with the required acumulation of masses of quantitative data to be made available for ecosystems analysis there is the necessary subsequent development of simulation models to be used, hopefully, in predicting plant disease, insect, nematode, and other outbreaks and epidemics. The development of predictive models is the goal of all ecosystems research and is emphasized by Apple (1974):

Development of ecosystems models must be the ultimate objective for all economically important agro-ecosystems. The development of an ecosystem model will require linkages between many submodels such as a plant growth model, pest complex model or models and biometeorological models. These are complex processes requiring extensive biological research and sophisticated mathematical techniques but these models can also be applied in assessing crop losses due to pests, which is a critical need in establishing researchable problem priorities. Although many discount the importance of modeling to a pest management system, I consider them essential. Not only do they serve as management tools but they dictate a systematic approach to pest management research.

Hooker (1975) emphasizes a systems approach in agricultural research. The need for backup systems in case of disaster in the primary systems is also emphasized:

> In many instances a systems approach, not a single control method, is required. Backup systems should be available and are needed in agriculture just as they are in aerospace programs. Normal cytoplasm provided a backup system for southern corn leaf blight (epidemic of 1970). In other diseases and crops we are not so fortunate.

Hooker (1975) points out that we tend to look backward and attempt to solve last year's disease and pest problems. Instead, we need many more efforts to think forward toward research programs designed to anticipate new problems that are arising or to keep particular problems from arising. In particular, efforts are needed to keep specific pathogens and pests from developing to epidemic proportions. This is the type of research that has received very little attention. It should receive much more in the future now that systems analyses can be attempted creatively and it must be done as an interdisciplinary effort.

One of the very difficult and most important problems in a systems approach to plant protection research is the careful examination and determination of the key variables in a system. These must be separated from the many variables that have a minor or insignificant role (Campbell 1972). Prior to computers the approach usually used involved the applied, adaptive research, field plot techniques that do eventually determine important *key* variables. These techniques operate very efficiently in a given situation on an empirical basis, but they are not able to help in the solution of certain other problems. The best laid plans have been altered and subverted by such problems as persistence of pesticides in the environment, other environmental pollution, the breakdown of plant resistance to a given disease or pest, the development of pest resistance to pesticides, the increased virulence to plant disease organisms, and the buildup of residue and toxicity problems associated with the utilization of certain agricultural chemicals. These problems have been neglected in the past simply because they have been too complex for the available research technology. Now, using the new systems approach, the computer and an interdisciplinary team and—where needed—the most complicated types of biochemical analyses, these problems can be faced creatively. They now must be considered along with the many other factors involved in any agro-ecosystem.

There are some who are highly enamored with the potential of systems analysis and who feel that the time honored, applied, adaptive research techniques associated with careful statistical field plot analyses are out of date. Nothing could be farther from the truth. The field plot techniques will be necessary to give much of the necessary ground truth data and to check the simulation models of the future against reality. There is no way to carry out a check on some computer analyses other than through careful, adaptive, field research techniques.

Also, in most of the world, computer technology is not going to be available in the foreseeable future. For this reason alone the adaptive field plot techniques aimed at understanding the effects of local environments on crop plants will be important. In fact, in most of the developing world the statistical field plot techniques that have been so helpful in the development of the outstanding plants of the "green revolution" will be essential in solving future problems in local areas. In spite of the known limitations of field plot techniques, and in spite of the fact that the information obtained is often only empirical, the results are so impressive that this type of research must be emphasized much more than it has been in the past. This point of view has been emphasized most eloquently by Borlaug (1972) as well as by many others.

Another area of interdisciplinary research needs to be emphasized much more than in the past, particularly in the developing world. This is the general area of harvesting, shipping, and storage losses. It has been variously estimated that from one-third to two-thirds of all food produced on earth is destroyed or consumed by plant disease organisms or various pests. At times the loss estimates for India, for example, go as high as 75 percent, and even in the most developed countries the total estimated losses are seldom under 25 percent (Brady 1974). In the developing world this is not only one of the major food problems but also one that could be solvable in the foreseeable future. It simply has not received the effort it has deserved. Bunting (1972) emphasizes the fact that these losses from various storage pests, the insects, the fungi, and the small mammals, are not only very expensive but discouraging and unnecessary. He suggests that the most rewarding research for plant protection in the near future in the developing world is to be found working on the problems of harvest, storage, and shipping. Brady (1974) also emphasizes this:

> The quickest gains in reducing food losses can be made by controlling storage wastes. Conservatively, two billion dollars worth of grain is lost each year in storage and transit.

Scientists from 27 nations attended a meeting on stored products entomology in Savannah, Georgia in October of 1974. They sent a resolution to the United Nations that emphasized the need for assistance of national and international leaders to encourage the utilization of available methods for the control of storage pests. The scientists emphasized that the comprehensive adoption of known technology could make a major contribution in solving worldwide shortages.

Most food in the developing world is not stored by

governments but by individual farmers in small villages under very poor conditions. In India, about 90 percent of all food grain is stored on farms or in villages in either jute sacks or more often simply piled in the corner of a building. The development of simple methods like sealing grain in clay pots, burying it in hollow trees, or the use of elevated grain cribs with rat proof shields on the legs would be a major improvement throughout Africa and Asia (Brady 1974).

Another area of interdisciplinary research that has been neglected in the immediate past and that deserves much more attention is that aimed at cultural controls of pests and plant diseases, particularly cultural techniques that for one reason or another alter populations drastically. In the more distant past, before the days of modern chemical control, such research was considered respectable. In recent years this research has been neglected. Actually, some of the cheapest and best control techniques have been the result of cultural controls. When combined in a systems approach with other control techniques, such as plant resistance and various biological controls, cultural controls may be the heart of effective control programs for many pests and diseases.

Certainly, the simplest and most effective method of plant protection for the grower to use, and probably the one most appreciated by farmers all over the world, is the use of resistant or immune crop varieties. Plant resistance to diseases and pests is fortunately the norm rather than the exception. Most plants are resistant or immune to most insects and diseases. It is only the exceptional insect or disease organism that causes the outstanding, economically significant epidemics. Were this not the case, man would have little hope of producing healthy crops. Since the use of a resistant or immune variety is simple and automatic, administrators and the public and sometimes even the growers tend to forget that breeding programs need to be supported constantly. Organisms are versatile and adaptive, and pests and disease organisms that can adapt and attack newly developed resistant varieties gradually but consistently appear on the scene. This is particularly true if the resistance bred into a given plant is controlled by a single gene. The search for outstanding multigenic resistance or, hopefully, immunity, needs much more support than it has received in the past, *and it is essentially an interdisciplinary activity.*

A related type of research that also needs much more support is that associated with the biochemical nature of resistance. Support for more research on all types of antagonisms is needed. Toxins, allelopathic substances, pheromones, phytoalexins, and all other biochemical materials produced by plants or microorganisms as mechanisms of antagonism or synergism need to be examined very carefully in relation to their possible role in plant resistance and immunity. Ultimately the biochemical nature not only of resistance but also of immunity needs to be understood. When obtained, this will be the most important possible information for plant protection. Eventually it may be possible to spray or treat a plant with chemical material that renders it highly resistant or immune to a given disease organism or pest. Here, too, we have a type of research that not only requires much interdisciplinary effort but that also may take so much cooperative time that frequent publications cannot be expected.

Unfortunately, when plant resistance is achieved the support for the research tends to be dropped. By contrast, more dramatic types of research such as spray programs that utilize large expensive equipment, even though they may give poorer control, tend to get continuous support. Administrators must understand that breeding programs must be long range and supported continuously over an indefinite period of time. If this approach is taken, then plant breeding in the long run will be the cheapest, simplest, and most useful method of plant protection, and certainly the one most desired by the farmer.

There has been much recent emphasis on the need for developing selective pesticides, essentially narrow spectrum chemicals that are effective against a single pest or a very small group of related pests and which do not kill or disturb desirable insects or other species of plants and animals. These selective chemicals will also need to denature quickly or at least must be shown to be innocuous in the environment to man and other warm-blooded animals. Unfortunately, it is difficult to convince private companies to go into the research and development of selective chemicals. The cost of research and development on a given pesticide is astronomical. One of the reasons the broad spectrum pesticides have been popular is that they could be expected to be used on a great many different pests, or in the case of fungicides on many different plant diseases. The selective narrow spectrum pesticides, by contrast, would be utilized only on specific pests or diseases on specific crops, and the revenue obtained by the private company developing the product would be much less. For this reason alone most chemical companies have been reluctant to initiate efforts in the development of selective chemicals.

This problem is discussed by Carlson and Castle (1972):

> Subsidies could be paid by government to chemical industries to help in the development of narrow spectrum chemicals. R. F. Smith and others have argued that there are many pesticides on the laboratory shelves that have been screened out because they would kill only a few species. An arrangement might be worked out in which university laboratories do the pharmacological studies and the patent is given to a private firm to produce and distribute the chemical, as is now done with hybrid research.

Certainly, the problem associated with the cost of research and development of selective pesticides needs to be faced more realistically than it has been in the past. Until some satisfactory, less costly arrangement can be made with private companies that makes it possible for them to develop these products at a profit, we cannot expect much progress in the development of selective pesticides. Some sort of interdisciplinary effort that includes public agencies as well as private companies would seem to be needed.

Another very important area of chemical research is the development of more combination pesticide programs. Whenever possible fungicides and insecticides should be combined in compatible combinations. Herbicides, nematicides and soil fungicides also should be combined. It is not inconceivable that eventually all necessary chemical treatments for a crop year—for all pests and diseases—might be applied in a single treatment. Problems of compatibility and phytotoxicity are very complex and this research must be developed across interdisciplinary lines. Certainly, the utilization of combinations of pesticides needs much more careful examination and research than it has received in the past.

A related subject that is becoming more important each year is environmental monitoring. It is increasingly necessary to check all residues, to make analyses for phytotoxicity and whatever other toxicological analyses are necessary after the utilization of given pesticides. Since all of these monitoring procedures are chemical or biochemical in nature, what is needed is the development of reasonably large interdisciplinary research laboratories that concentrate on the problems of residues, tolerances, and the like. It is usually a mistake to develop laboratories for analysis of only one type or even a few types of toxicological material. A much more realistic approach for any institution is to combine this whole chemical and biochemical effort in a single interdisciplinary laboratory that is well-staffed and equipped.

All interdisciplinary plant protection research needs to be coordinated with agronomic and horticultural research, *and be subordinate to it.* As we have stressed earlier, the most important points to consider always are those associated with the improvement, production, and management of a specific crop. Plant protection must always be an important but subordinate part of this overall effort. It has been extremely difficult to convince the separate disciplinary groups that this philosophy is essential. The disciplines have tended to be parochial and defensive about this matter. Probably for this reason the outstanding applied interdisciplinary research in crops usually has not been done by educational research institutions but by the large international foundation-supported research institutions and by large commercial plantation organizations such as those growing sugar cane, pineapple, bananas, coffee, tea, and so forth.

There are some exceptions to this observation, however.

For example, in the United States there have been outstanding voluntary programs of cooperative interdisciplinary research in which the Department of Agriculture has cooperated with various states in the development of improved corn, wheat, and other small grain varieties. These crop improvement associations have been very productive in the past and have been the source of the "green revolution" which occurred in the United States and in other technologically advanced countries many years prior to the "green revolution" of the developing world which has been given so much publicity recently.

It is extremely important to organize interdisciplinary plant protection research around a crop, or perhaps a crop rotational system, and include in the research team all the necessary, relevant disciplines. These must be controlled by agriculturally trained administrators who understand and appreciate the importance of interdisciplinary research as applied to the solution of field problems in specific crops. The funds for such research need to be specifically designated for cooperative, coordinated, interdisciplinary research projects that continue for an indefinite period and from which no immediate publications are expected. Promotions must be geared somehow to the success of the team rather than to the productivity of each individual and they also must be geared to the solution of problems rather than the publication of papers or groups of papers. In short, the prestige, power, pay, and all other perquisites involved with this type of research must be as great and as advantageous as those given to the individual working in basic research.

Sherf (1973) describes the ideal interdisciplinary, cooperative, applied research team. Such research teams, designed to solve problems in a given crop, or a rotational crop system, need to be organized from top to bottom reaching all the way across to the farmer. The basic or more sophisticated research personnel need to be organized in interdisciplinary research teams. Working with these teams should be the upper level extension or applied research specialists, those trained in applied, adaptive research methods that will test the results of the first research team. Working with the applied research specialists would be the various regional specialists in each discipline, the well-trained industrial representatives and/or the field extension specialists. These would not only work with those persons involved in applied, adaptive research, but also sometimes they actually would be involved in the research program at the regional level. They also would be contacting growers in groups or individually. Working with this group would be some very important people usually missing in the developing world and often missing in the developed world. These would include the County Agricultural Agent, the private practitioner in plant protection, the generalist who functions as a pesticide specialist either for the extension service or through a chemical company, the chemical salesman, the seedsman, the field scout, and so forth. Not to be forgotten as an integral part of a successful applied, adaptive research system is the grower himself, whether he be a large operator or small. In order to function well the system needs to get research results and information to the field men very rapidly and also get the field problems to the research personnel with equal rapidity. This rapid two-way communication is extremely important in the success of interdisciplinary research programs designed to solve real crop field problems. In charge of this effort must be a very well trained agricultural administrator who is sympathetic to the total goals of the program, who realizes the length of time required for this type of adaptive research, and who also understands that it cannot be handled in the same way that the basic individual research programs are usually handled (Sherf 1973).

Many outstanding individuals were contacted by the author concerning their attitudes toward interdisciplinary, cooperative, applied research programs, and they were asked to suggest research organizational patterns. The pattern given below by Schmidt (1975) is the most popular sequence of choices.

> My first choice would be the team research approach by problem areas quite apart from disciplines as being the most effective method for reconciling basic differences between the disciplines. My second choice would be the pattern of informal cooperative

> research similar to that utilized in the Hard-Red Winter Wheat Improvement Program in the central U.S.A. My third choice would be the assignment of individuals by departments to work on common interdisciplinary research problems in a research team.

Schmidt's first choice, it is interesting to note, is essentially the same organization as that just outlined above by Sherf and basically the pattern of research organization that produced the recent "green revolution."

CONFLICTS IN CONTROL RECOMMENDATIONS

There are a great many conflicts between field recommendations given by agronomists and horticulturists and those given by plant protection specialists. There are also many conflicts in recommendations between such fields as entomology and plant pathology. Reconciliation of these conflicts and the achievement of satisfactory compromises has always been neglected; in fact, it is almost never discussed.

A few examples of such conflicts follow. Many others could be cited.

The European corn borer has been controlled in the past by the destruction, or else farm use, of all crop residues plus deep mold-board plowing. The recent development of minimum tillage practices in the United States works against these control measures.

One of the recommended methods for grasshopper control in the United States has been the destruction of eggs in the fall and winter by deep plowing thereby exposing the eggs to harsh weather and to birds. Minimum tillage also works against this practice (Metcalf *et al.* 1951).

Sometimes unwise laws conflict with pest control recommendations. For example, in California a law has been passed that allows the growth of only a single cotton varietal type. Over the years this has led to very severe pest problems. Had rotation of different varieties been encouraged, particularly varieties having varying patterns of resistance to the various pests, this problem would have been much less severe (Hooker 1975).

Quite a few conflicts occur because of plant breeding changes. Often plants with a very high level of resistance, perhaps even immunity, have poor or unacceptable quality or other undesirable agronomic characteristics. Certainly, the breeding of plants resistant to pests and diseases is replete with examples of success in controlling one pest only to find that the plant is very susceptible to other equally important pests (McKelvey 1975).

Plant pathologists have recommended low plant populations in corn for stalk-rot control. Agronomists, by contrast, have recommended high plant populations for best yields in hybrid corn production. A compromise recommendation has gradually evolved which really is an intermediate plant population (Schmidt 1975).

Apple (1972) emphasized some of these conflicts and pointed out the necessity for reconciling and compromising control recommendations at the field level so that conflicts in recommendations can either be minimized or eliminated.

> Many management practices of modern agriculture enhance susceptibility to disease or attack by insects. These practices include:
>
> 1) Fertilization which produces larger and more succulent plants that are often more susceptible to disease or insect damage than plants grown at lower nutritional levels.
> 2) Irrigation which favors many disease and insect pests as contrasted to the fluctuating soil moisture levels under natural rainfall conditions.
> 3) Tillage which Yarwood (1968) found to be an important factor in increasing the incidence of disease as compared to no tillage culture.
> 4) Multiple cropping which promotes rapid build-up of pest populations.
> 5) High plant population densities which change the microenvironment which favors the development of some pests.

Many more examples could be given but strange as it may seem, very few have been documented or published. Each well-trained, experienced field research worker has his own collection of recommendation conflicts which he passes on to friends but seldom discusses elsewhere and apparently never publishes if it can be avoided.

Without question much more research needs to be done to resolve these conflicts. In the long run, however, the recommendations of the plant protection people must be adapted to the primary needs of agronomy and horticulture. The most important factor always is the production of a high-yield, high-quality, healthy crop and all disease and pest control procedures must be subservient to that goal.

EFFORTS TO DEVELOP COOPERATIVE RESEARCH PROGRAMS AND NETWORKS

Informal, voluntary, interdisciplinary, cooperative research programs are often more successful than formal programs. Probably this is true because the informal efforts are usually staffed by people who are determined and willing to be cooperative and unselfish in their approach toward the solution of what they consider to be important problems. Formal organizations often bog down because the individuals involved are not cooperative or subservient to the larger goals, and they are primarily concerned with personal matters of pay, prestige, promotion, publication, control of personnel, equipment, and the like. As discussed earlier, the problem of "human cussedness" seems to be the most difficult of all problems to solve.

Prior to World War II most of the outstanding cooperative research in plant protection and pest management in the developing world was done by the large plantation industries on banana, coffee, tea, sugar cane, pineapple, and the like. The millions of small farmers of the developing world have always been a neglected and forgotten group. Since World War II, there has been the gradual development of international research centers on several crops. These have been co-sponsored largely by international foundations such as the Rockefeller and Ford Foundations plus the local governments where the institutions are located. The outstanding research center in Mexico on wheat and corn and in the Philippines on rice have been present longer than most of these institutions and have received the most publicity. However, in recent years a foundation-sponsored research center on sorghum and several other crops has been initiated near Hyderabad, India, and one on vegetables in southern Taiwan. The International Institute of Tropical Agriculture near Ibadan, Nigeria, concentrates on several major African crops. These institutions have done and are doing brilliant research on specific crops, and the recent publicity concerning the "green revolution," particularly in Asia and earlier in Mexico, is largely an outgrowth of the brilliant basic and applied research done at the foundation-sponsored international research centers.

In several countries of the developing world outstanding in-country agricultural research institutions have been initiated. A good example of one of these would be the Indian Agricultural Research Institute located near New Delhi. There are several others that have done and are doing outstanding research in specific countries.

Plant diseases and pests move easily across country boundaries making their control an international problem. Consequently, there has been a recent push by individuals as well as organizations to encourage the development of coordinated, cooperative, interdisciplinary, international research networks in agriculture and specifically in plant protection. In 1971 President Richard Nixon suggested such a new approach for the US/AID program (Apple 1974). He suggested an International Development Institute that would be carried out through the establishment of a coordinated international research network. Eminent, internationally oriented plant pathologists such as Thurston (1974) have supported such a program. Thurston suggested a worldwide cooperative effort to monitor pathogens and perhaps other pests, and he has recommended the example of scientists and governments in Southeast Asia in their recent successful effort to prevent the introduction of South American leaf blight of rubber as being a fine model of how such a cooperative effort might work (Thurston 1974).

He suggested that the various national phytopathological societies and national and international research institutions need to develop a system of coordinated study of various pathogens on a worldwide basis and develop cooperative information exchange systems, seed banks, and gene pools for crop breeders. He emphasized also that studies on etiology and epidemiology of many of the tropical pathogens are completely inadequate and are needed now. In particular, personnel centers or pools composed of well-trained people interested in working on international plant disease problems are needed.

Apple (1971) also has discussed the subject of an international research network for plant protection.

> The major rationale for an international research network that focuses on pest management problems is the fact that many of the most serious pest problems of the world's major food crops are world wide in distribution and importance. Such pest problem complexes must be studied on an international scale if they are to be understood so that ecologically and economically sound, long term management programs can be developed. Pests, whether they be insects, pathogenic microorganisms or noxious weeds neither know nor respect national or continental boundaries. There is a great need to have an international perspective of many important pest problems because they are location specific. Varieties may be formed differently between areas because of differences in races or strains of a pathogen, or endemic organisms which do not cause economic damage on indigenous varieties may be highly destructive on newly introduced types, or insects that are not a problem in one area may be destructive in another because of differences in cultural practices, absences or presence of natural predators, and so forth: or the environmental conditions may influence the effectiveness of a pesticide because of differences in pH, temperature, humidity or many other factors. All of these and many more location specific attributes of pest problems support the need for an international research network for pest management to provide greater stability for the agriculture in the developed and developing worlds. The international research network would not be in competition with the international (research) institutes, rather it would serve to maximize the utilization of their results. The network would focus on country level problems but with regional overview so that communication between individual scientists working on similar problems is maximized and duplication of effort is minimized.
>
> An important function of the network will be to coordinate project activities with related programs of other agencies, foun-

dations and governments. One objective would be to have the pest management research network serve as a communications medium between the many independent program activities currently active in the developing nations. Cooperation or liaison would be established at the international research centers (for example CIAT, CIMMYT, IRRI, IITA and others), international development agencies (for example World Bank, FAO, UNDP, USAID), private foundations, bilateral country programs and in-country research programs with regional or international significance. Coordination would be enhanced by the formation of an informal "international pest management group."

Apple also has suggested that the International Weed Control Program initiated a few years ago at Oregon State University under US/AID funding might be the first step in the development of such an international effort.

Certainly, there are many differences of opinion concerning how to go about developing a meaningful international, coordinated research effort in plant protection. However, there would be few who would not recognize the need for such an effort. Probably it would be wise to utilize international auspices, perhaps through the United Nations. An FAO-centered international program with a center of communications as well as a center for research information coordination would have a much better chance to be accepted and become effective in the developing world than one sponsored unilaterally, for example, by institutions in the United States or any other western country. It would be wise to have a series of exploratory meetings under international auspices to study the possibility of developing such an international research network.

Chiu and Yen (1972) have discussed the needs of the developing world in plant protection and also have encouraged the development of international programs to help solve these needs. They suggest that practical research and training programs are needed at all levels. They feel that training programs, and not just research programs, are essential if long-term improvement in plant protection is to be achieved in the developing world. They list the following goals:

1) A reduced reliance on pesticides, especially with the low-value per acre food and feed crops;

2) The gradual elimination of non-degradable or slowly denatured pesticides so that environmental and residue hazards can be minimized;

3) A sharply increased reliance on plant breeding and genetic means of pest population suppression;

4) An integration through coordinated control effort of all measures that contribute to the achievement of a low or economically acceptable level of pest populations utilizing techniques that can be used successfully in mechanized crop cultivation as well as primitive cultivation.

In each case economic factors need to be considered very carefully. It should be emphasized that these four needs for the developing world are also essentially the needs of the developed world. No matter what sort of coordinated international effort in research (and training) is eventually developed, the most important concept to remember is the absolute necessity of well-planned adaptive research in all matters associated with plant protection. These adaptive research techniques centered around the time-honored statistical field plots are the best hope of achieving accurate applied research data which can be used practically in local situations. The empirical results achieved through randomized and replicated trials over many years in each of the different climatic zones and on all of the different soil types are absolutely essential, and no sophisticated computer techniques are going to eliminate the necessity of utilizing these other techniques to determine the most acceptable varieties, cultural methods, pesticides, and so forth for a given localized area (Borlaug 1972).

There has been a trend in recent years for many in agriculture, particularly agricultural administrators, to belittle the field plot techniques. This trend must be reversed and it must be understood that there is no other known way at this time to let nature tell us in an empirical fashion what will work best in a local situation. This is particularly true in the developing world where sophisticated computer techniques or simulation modeling techiques will be available only in isolated instances, at least in the foreseeable future. But this is also equally true in the developed world where the sophisticated computer techniques have to be checked constantly with the ground truth data of the real world at the field level.

EXAMPLES OF COOPERATIVE RESEARCH FROM THE UNITED STATES

In the United States there are examples of cooperative, coordinated, interdisciplinary research in plant protection that go back many years. Probably the oldest attempt to develop a cooperative program of any type in insect control was started in Arkansas (Lincoln *et al.* 1963). A boll weevil scouting program was started in 1925 with one scout, James Horsfall, now an eminent plant pathologist. Over the years it has increased to include agronomic and cultural information, weather and environmental effects, possible biological control, and insecticide recommendations. Most recently, weed and disease information has been included and some recommendations have evolved from this information. At present, there are about 150 scouts working on the cotton crop in Arkansas. These are for the most part temporary summer workers and are trained at the beginning of each season to work specifically on the cotton crop that year.

In several of the eastern states there is a long history of research efforts to combine insecticides and fungicides in spray schedules on such crops as apples, peaches, pears, potatoes, and some of the other fruit and vegetable crops requiring rather complicated spray schedules. This cooperation has usually been somewhat informal between departments of plant pathology and entomology in the various state land-grant institutions.

There are many examples of informal cooperation between individual agronomists, plant pathologists, and entomologists. These could probably be found on most of the land-grant campuses in the country. An example of one given by Schmidt (1975) would be the informal cooperation between the plant pathology and entomology departments at the University of Nebraska in their studies on wheat streak mosaic virus. This would also be true in Kansas where the largely informal cooperation between agronomists, entomologists, and plant pathologists studying wheat streak mosaic virus and other small grain viruses has been outstanding.

In Arkansas the disease and pest problems of the peach crop have been approached cooperatively by horticulturists, entomologists, and plant pathologists for many years (Dickerson 1974). Again, in Kansas, cooperative research has been characteristic of most of the work done on problems of small grains such as rusts, viruses, the Hessian fly, and various other insect problems, particularly those involved as virus vectors. This cooperative program has always included plant breeding and has always included the critically important agronomic characteristics of the crop such as yield, vigor, and quality (Dickerson 1974).

Many other similar cases could be given from other states. For example, in Nebraska (Schmidt 1975) the agronomy and plant pathology departments have cooperated on the soil-borne wheat mosaic virus problem; agronomy and entomology have cooperated on the corn root worm problem; and agronomy and plant pathology have done cooperative work on freckles and wilt on corn. Schmidt suggests that there have been no discipline problems encountered here and this work is distinguished by very fine informal cooperative efforts. Some feel that this is the best and most meaningful type of cooperation and that formal cooperation quite often breaks down because of personality and other human factors that are very difficult and often impossible to control in a local situation.

At the University of Wisconsin in the Department of Plant Pathology there are at present 25 formal cooperative projects with other departments (Kelman 1974). There are some in which two or three departments are involved, including plant pathology. Cooperative projects with plant pathology include the following departments: Horticulture 9, Agronomy 8, Soils 2, Entomology 3, Food Science 2, Agricultural Engineering 1, Forestry 1, Veterinary Science 1, and Dairy

Science 1. These cooperative efforts, according to Kelman, have resulted in varying degrees of control of plant diseases. Not all, of course, have resulted in complete success but progress has been made in most cases.

Similar examples of formal cooperative efforts in other states between various departments in the plant protection disciplines can be given. For example, two three-year grants to support a pilot project in integrated pest management were recently awarded to seven scientists from three different departments at Michigan State University (Anon. 1975f).

There are also formal cooperative contracts between federal and state personnel. Most of these are initiated through the U.S. Department of Agriculture and are supported by federal funds. However, many of these also have been essentially informal cooperative efforts, for example, the long-time, largely informal cooperative efforts characteristic of such organizations as the Hard Red Winter Wheat Improvement Association. Similar efforts have been present in the soft red winter wheat area of the United States, the spring wheat area, and on corn in several corn belt states.

In 1970 there was an outstanding example of a combination of informal and formal research cooperation as a result of the advent of the southern corn leaf blight in the northern corn belt of the United States (Hooker 1975; Schmidt 1975). Research personnel from the federal and state governments, universities, and industry representatives met on several occasions and outlined a plan of attack to solve the problem quickly. In Nebraska, a cooperative experiment station committee was set up immediately with representatives from plant pathology and agronomy. This was done in cooperation with other states in which the southern corn leaf blight outbreak occurred. In this particular crisis, no interdisciplinary conflicts were encountered and all worked feverishly in a cooperative effort to solve what was an extremely serious problem (Schmidt 1975). Usually in extreme crises such as this the clashes between the disciplines are forgotten quickly and all get together in a massive effort to solve a problem. This has been characteristic of cooperative efforts in the United States in crisis situations on other occasions and there are many who are firmly convinced that this is still the best way to achieve successful cooperative efforts between the disciplines.

Recently, more formal, large-scale efforts have been initiated in developing cooperative research efforts in plant protection. In 1972 a U.S. Department of Agriculture Interagency Pest Management Working Group was established that is developing cooperative programs with the states and also with industry (Robins 1972). Several of the department's agencies participate: the Agricultural Research Service, the Forest Service, the Cooperative State Research Service, the Animal and Plant Health Service, the Extension Service, Economic Research Service, and the Office of Information. This interagency group provides coordination, communication, and advice on program activities, and it furnishes advice and information to the Director of Science and Education. A research subgroup was established that included the Agricultural Research Service, Cooperative State Research Service, and the Forest Service. This group represents the federal government in insect research program development in the states, the National Science Foundation, and the International Biological Program. It is hoped that the program will be expanded soon and plans are being made to include weed, nematode, and disease research interests.

Also, four regional subgroups have been formed to advise on the pilot programs. These include the various state experiment stations, the state departments of agriculture, and the state extension services. Also included in the regional groups are the U.S. Department of Agriculture Extension Service, the Agricultural Research Service, the Animal and Plant Health Service, and the Cooperative State Research Service. A steering committee for the whole program has been formed to give advice on the development of cooperative research programs and includes personnel from the U.S. Department of Agriculture, the Environmental Protection Agency, and the National Science Foundation.

It has been recommended that states form similar organizations and develop lines of communications with the federally sponsored groups and that they include the states' departments of agriculture, the Agricultural Experiment Station, the Cooperative Extension Service, and industrial and growers' groups. Special emphasis is being placed on the effort to develop communications and to gain input from the chemical industry and from various growers' groups (Robins 1972).

In 1972 an integrated cotton insect management project was started by the U.S. Department of Agriculture Extension Service in 14 states (Good 1974b). USDA funds were furnished,$2 million to be utilized over a period of about three years. Initially, this was to be an integrated insect control program including agronomic practices. The intended goals follow:

(1) The control practices achieved must be economically practical and at least cost no more (and preferably less) than earlier control efforts.

(2) The insect control must be as good or better than in the past, and the yield and quality of the crop must be equal to or better than in the past.

(3) The control procedures developed must be completely compatible with agronomic and other pest management practices and must not affect the local environment adversely.

(4) All federal and state environmental regulations must be followed.

(5) Growers involved must help plan and develop all the projects and must participate through their various organizations.

(6) In all cases residue analyses must be made and environmental monitoring must be included.

(7) All workers must be checked regularly by local health departments, particularly those involved with the use of pesticides. It is hoped that this plan can be developed to include weed, nematode, and plant disease problems on cotton in the near future (Good 1974b).

Also in 1972 and 1973, 39 pilot pest management projects were established in 29 states on 15 commodities sponsored by the U.S. Department of Agriculture Extension Service. This group of pilot pest management projects was to extend for a period of three years, testing and adapting technologies of crop protection and pest management to local or regional field situations (Good 1974a). Good's description of this effort follows:

> These projects were implemented because sufficient technology was available to begin the development of integrated pest management through large demonstration programs which require grower participation and support. As now constituted the pilot projects deal with the following crops and pests; fourteen projects deal with management of cotton insects; six with insects and weeds on corn; four with insects and weeds on grain sorghum; two with insects, diseases, weeds and nematodes on peanuts; six with pest complexes on apples, pears, and citrus; four with insects on vegetables and potatoes; two with management of insects on alfalfa; and a single project in 1974 will attempt integrated management of insects, diseases, weeds, nematodes and sucker control on tobacco.

These projects are built around voluntary grower acceptance and support of the programs. Actually the growers are expected to gradually take over the funding of these programs if more reasonable and effective control is achieved. The success of these programs has been considerable since 1972, particularly in the 14 cotton projects. This success probably is related to the long years of experience in cotton insect scouting in most of the states and the fact that the system had been developed earlier to the point where it was possible to expand to larger cotton acreages. Growers have supported and are contributing heavily to the program. They contributed $458,000 in 1972, $822,000 in 1973, and $1.2 million was projected for 1974 on 860,000 acres of cotton. The growers are paying all the direct costs of the insect scouting services in 11 out of the 14 states involved and are furnishing approximately half of the funds being utilized. The net returns over conventional insect control methods have been reported as high as $95 per acre.

Wherever possible an effort is being made to include plant disease, nematode, and weed problems in these programs. There is also an effort to minimize the use of pesticides and move to non-chemical control techniques for the regulation of pest populations, techniques such as crop destruction, management of pests in surrounding nearby vegetation and crops, destruction of sources of pest infestations both inside and outside the pest management areas, and crop destruction and management of ground cover (Good 1974c).

Good summarizes this program and the hopes for it in the following statement:

> The state pilot pest management programs now under way and those to be sponsored in the future by the US Department of Agriculture must accommodate in a harmonious system the entomologist's concept of insect management, the plant pathologist's concept of plant health, the nematologist's concepts of population regulation, and the weed scientist's views on total weed management on farms. This is necessary because management of farm units requires a systems approach to controlling a large array of pests, all of which can contribute to reduced yields and quality. Failure to control a single component of the pest complex often negates the benefits of other controls. In like manner control of one pest often has an inescapable effect on other pests. The objective of the pilot pest projects are to establish multiple and alternate choice systems of pest control that are effective, economical and environmentally sound. The ultimate goal of these projects is to promote effective use of combinations of cultural, biological, and chemical methods.

Certainly, few would quarrel with the hopes and goals for these projects. It is hoped that a more even-handed approach can be taken in the future than often has been apparent in the past so that each of the plant protection disciplines will be considered as complete equals in the planning and execution of the total research effort. Heretofore, these projects have been largely dominated by entomologists. This must change if other disciplines are to enter these projects with equivalent enthusiasm.

Recently, as part of the International Biological Program, 19 universities, parts of the U.S. Department of Agriculture, Agricultural Research Service, the Forest Service, and some elements from private industry have joined together in an interdisciplinary coordinated research attempt to seek some practical alternatives to the large-scale use of the broad-spectrum pesticides for the control of certain insect pests and pest complexes. Six major crops were picked: soybeans, cotton, alfalfa, citrus, pine (bark beetle), and pome and stone fruits. Apple (1974) thinks that this research activity is the most important pest management research activity now being funded by the National Science Foundation and the Environmental Protection Agency. It is under the direction of Carl Huffaker and Ray F. Smith of the University of California, Berkeley, and is often referred to as the Huffaker Project.

Although conceived initially as an interdisciplinary research project, it has actually developed largely as an integrated effort in the control of certain insect pests. There are a few pathologists, nematologists, agronomists, horticulturalists, and weed science personnel listed in the master project proposal. Some of these actually have been utilized in an attempt to develop interdisciplinary research programs. It is clear in the first year's report, however, that this research program is dominated by entomologists and the thinking and the point of view is largely entomological (Huffaker 1974).

All of the plant pathologists and nematologists listed on the project and some of the weed scientists were contacted and most replied concerning their participation or nonparticipation in the project. One pathologist who was listed in one of the states had never heard of the Huffaker Project even though he was listed as the plant pathology collaborator for the state. He went to the Department of Entomology to check on the project and found that five entomologists were involved and knew a lot about the project, but he had never even heard of it. Another plant pathologist reported that the entire program in his state was slanted toward integrated insect control. He commented as follows:

> I happen to be involved by being a member of the so-called steering committee. I think of it as an advisory committee. I suppose in the light of the present day concern about insecticides polluting the environment, this type of research had the better chance of being funded at the time of the proposal, but nothing in our program thus far concerns plant pathology at all.

Another plant pathologist reported that the plant pathologists were being left out of the project completely but that recently great pressure was brought to bear on the Huffaker group to make room available for plant pathology research in the program. He comments on the results of that effort as follows:

> Since this meeting we have been contacted by representatives of the Huffaker Integrated Pest Management Project and have been invited to submit a new grant proposal in our areas of interest.

Without question the project is an outstanding one just from the standpoint of developing better methods of integrated insect control. In defense of the project, Good (1975), who is a nematologist and is serving on the National Steering Committee representing the U.S. Department of Agriculture Pest Management Program, has the following to say:

> In 1971 and 1972, when the Huffaker project was being formulated, there was little interest in support of integrated pest management by other disciplines; therefore the entomologists proceeded to provide leadership in this area. They are to be commended for their foresight and now their willingness to broaden the base of integrated pest management.

Not many plant pathologists would agree with Good's statement since most are thoroughly convinced that they have been involved from the very earliest times with a coordinated type of plant protection equivalent to integrated pest management.

In its initial report, the Huffaker Project (Huffaker 1974) has included brief reports on three plant disease programs that are being combined with the insect control programs. Hence, it is clear that at least some effort is being made to include plant diseases in certain segments of the project.

The Huffaker Project is, despite its detractors and problems, an outstanding and honest effort to develop integrated control techniques for insects and, hopefully, eventually for some other pests and plant diseases. It does illustrate very well the difficulty inherent in the development of large formal interdisciplinary research programs and, actually, Huffaker and his co-workers should be highly commended for their efforts.

One of the most outstanding interdisciplinary research

efforts seen thus far has been developed by the New York State College of Agriculture in its effort to control all of the apple pests and disease problems in a single cooperative program (Arneson 1975). Here also there has been some problem with achieving cooperation in the project, but a reorganization of the whole program in 1974 seems to have solved this problem, and it would appear that this particular project has a good chance of including all disciplines as equals (Brann and Tette 1973). The project involves horticulturalists, entomologists, and plant pathologists, extension personnel, growers' groups, and industrial personnel. The table of organization (Table 12.1) is of considerable interest since it emphasizes the importance of an extremely strong and well-trained coordinator in the key post. This Coordinator is in complete charge of the program and works through a Manager of Operations who controls the field level research activities under the direction of the Coordinator. The Coordinator has an advisory group composed of growers, growers' cooperatives, extension agents, and industrial representatives. He also has separate advisory committees of research specialists in pomology, entomology, and plant pathology, plus several other disciplines including agricultural economics, natural resources, and meteorology. There are two steering committees who develop policy matters for the program and under whom the Coordinator functions. One steering committee is made up of representatives of various related state and federal agencies. Another one is made up of research directors, the director of extension, and the chairmen of the participating departments.

The table of organization is excellent, but for it to be effective an administrative device must be developed by which the different individuals and disciplines are given the pay, prestige, and perquisites necessary to keep them happy within the framework of an interdisciplinary, cooperative research program in which even the relatively few publications that are produced are cooperatively or jointly published. In spite of the best laid plans, such a program may collapse because of inadequate leadership or petty bickering among the disciplines and individuals. Most formal, cooperative, interdisciplinary projects have collapsed essentially

Table 12.1 New York State Pest Management Projects Table of Organization 1975

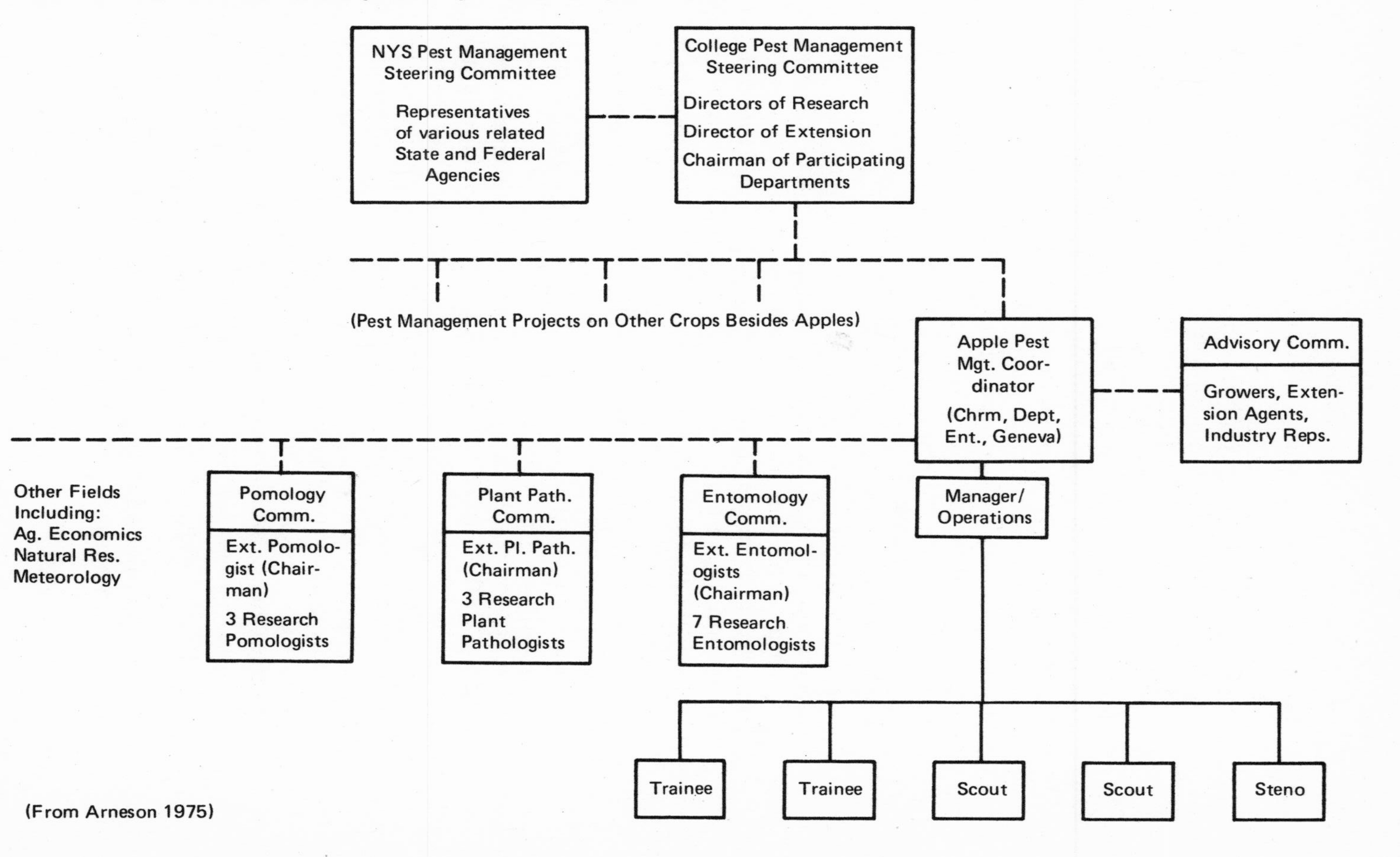

(From Arneson 1975)

because the participants felt they were not being treated fairly.

An interesting program has been developed at Oklahoma State University. This is a completely coordinated program aimed primarily at peanut production, improvement, and management. The overall program concentrates on peanut health. It includes disease, weed, insect, and nematode control, but it is subservient to the needs of peanut improvement, production, and management. A peanut production guide is distributed for the use of extension personnel and growers and also a brochure entitled "Peanut Health Program" which gives growers a list of just what to do during each month when the peanuts are being grown. A farm record is kept by each cooperating grower and is also kept by extension workers involved with the program. It includes a record of all the herbicides, fungicides, nematicides, and insecticides used. It also lists fertilizer use and all other chemicals plus any other cultural techniques that were involved in crop management and production. It includes yield and harvest and quality data. All of the relevant disciplines are involved in this cooperative program. These include agronomy, plant breeding, irrigation, agricultural engineering and, of course, plant pathology, nematology, weed science, and entomology.

This is primarily a state program but it is carried out in cooperation with the Southwestern Peanut Growers Association. It is one of the best overall cooperative programs seen thus far in the United States. In this case, several important rules have been followed that have helped in achieving success. The concentration of effort is on a specific crop. The primary emphasis is on crop improvement, production, and management. The secondary emphases are in other areas, including plant protection. These fulfill their proper secondary or subsidiary roles in the overall program of peanut improvement, production, and management (Sturgeon 1974a).

EXAMPLES OF COOPERATIVE RESEARCH OUTSIDE THE U.S.A.

An outstanding example of cooperative and coordinated effort in an agricultural industry (citrus) is reported from

Israel by Harpaz and Rosen (1971). The program is described as follows:

> The citrus industry of Israel is organized in a unique manner which renders it most suitable for development and successful implementation of novel pest control strategies based on integrated control principles. The Citrus Board of Israel is the sole organization through which all of the citrus crop of the country is marketed whether for export or local consumption. The Board is empowered to collect certain levies imposed on the proceeds of citrus fruit sales and through its agro-technical division the Board finances research and development projects covering all aspects of citrus production and marketing, among which pest control is the most prominent. The Board is authorized to carry out pest control operations on citrus on a country-wide scale. It also acts as the major purchaser and distributor of insecticides and pesticides on behalf of the country's growers for reasons both of economy and quality control.

The achievements of this organization are impressive, particularly in the area of pest control, and specifically in the biological control of the Florida red scale, *Chrysomphalus aonidum*. This type of organization would be extremely helpful in many small countries of the world and in certain industries in some of the larger countries. It is recommended as a case for serious study for those interested in developing a completely coordinated program of plant protection closely related to and organized under the umbrella of crop improvement, production, management, and marketing of a specific crop.

The Republic of China (Taiwan), another small country that has developed very rapidly in recent years and has become quite efficient in its agricultural production and marketing, has developed an Integrated Rice Production Program and also an Integrated Vegetable Production Program. This could be confused with the term integrated pest control as used in the United States, but actually it is more nearly the equivalent of the International Rice Research Institute's concept of a "Package of Practices," a concept also used in India and elsewhere. This is a well-coordinated crop improvement, production, and management program that also includes all aspects of plant protection (Luh 1971).

Also in Taiwan, on rice and several other crops, a very

closely coordinated cooperative effort in plant protection has been developed which includes plant disease forecasting. This is an outstanding example of each discipline being involved in the research efforts. The results are being applied very well at the field level with enthusiastic grower cooperation and participation. It, too, is completely tied in with the improvement, production, and management program in rice and the other crops involved. Without question, this is one of the most outstanding cooperative programs seen thus far and is to be greatly commended, particularly for use in the developing world (Chiu 1974).

In India, a coordinated multidisciplinary program of national research projects on different crops has been initiated. The main objective of these projects is to undertake research on various aspects of crop production, including plant protection, in different agro-climatic areas of the country. In these projects a number of research centers participate in conducting experiments in a coordinated manner. For each project there is a project coordinator who is the leader of the total research effort on a particular crop in a specific area of the country. There is a national coordinator for each crop or group of crops (Swarup 1974).

Appa Rao (1974) also emphasizes the importance of these operational research projects for integrated pest control in India. He describes coordinated, interdisciplinary projects that include from one to two thousand acres of contiguous area belonging to separate farmers growing essentially a single crop. The integrated pest control measures are practiced in a cooperative effort on these acres. All operations are done by the farmers under the supervision of the university scientists and extension workers from the Indian state departments of agriculture. Participating departments include plant pathology, entomology, agronomy, and plant breeding. Both arthropods and plant disease problems are included for each crop, along with fertilizer deficiency and toxicity problems, and other recommended crop production techniques.

These coordinated plant protection programs in India are

certainly a strong step in the right direction. They are aimed primarily at a given crop or a group of related crops. They have crop improvement and production as the primary concerns and include all of the relevant plant protection disciplines as part of the cooperative effort.

Chapter 13

Extension: Problems, Needs Strategies

GROWER ACCEPTANCE

The fundamental task of agricultural extension is that of carrying research information to the farmer or grower in a way that can be understood and used with convenience and economy: A most difficult task. In many instances this has been done remarkably well in the United States, and not so well in other instances. However, there always seems to be a considerable gap between actual available knowledge and practice at the field level. Closing that gap is the constant battle of extension personnel. Research information for a crop is often not useful to the grower until it is organized into a convenient "package of practices," a picturesque term used by the International Rice Research Institute in the Philippines and elsewhere in Asia. This "package of practices" must include all of the necessary economic and agronomic information as well as that on plant protection. The task of presenting this concise body of information either in visual, written, or verbal form is a difficult one, and it is particularly difficult in the developing world where most small farmers are still functionally illiterate and where relatively few even have access to the transistor radio. Also, when one remembers that there are few trained agricultural extension workers or organized extension programs in the developing world, the problems often seem insurmountable.

E. Smith (1972) summarizes the gap between knowledge and practice that still exists in developing countries:

> A tremendous gap exists between knowledge and its application in matters pertaining to the well-being of the individual. Herein lies some of the paradoxes of our time and the sources of greatest frustration. In the area of pest control some of our oldest strategies, proven sound by the test of time, have not been generally implemented. For a nation so committed to education, it would seem that we could do far better.

Interestingly, this statement was aimed at the United States which has had an outstanding Agricultural Extension Service dating back to the early 1900's.

It does little good to develop either sophisticated or simple control techniques for various plant protection problems if the growers will not accept them or forget to use them. As Good (1974c) wisely points out, "Grower acceptance is the most important factor in implementing pest management programs."

The grower must always be convinced that it is to his economic as well as agronomic advantage to use the control techniques, and these techniques must be made available in a simple, understandable form that is convenient to use.

NEED FOR BETTER TRAINED PERSONNEL

In recent years there has been a movement toward integrated pest management. This is creating a more sophisticated approach toward insect control than the methods usually recommended in the past. The future development of interdisciplinary, coordinated or combination control procedures for plant diseases as well as pests will in many cases result in an even more sophisticated approach and will require better trained personnel. This will compound the problem for agricultural extension personnel and mean not only more people working at all levels, but also better trained people. This need was emphasized some time ago (Anon. 1965):

> No matter how effective techniques of pest control may be, if they are to be adopted and used by growers, it will be necessary to educate extension specialists in their use. Effective use of integrated control will require the reorientation of almost all extension personnel to enable them to change from a relatively simple concept of single factor approach to the more sophisticated approach based on principles of applied ecology. This change is necessary if the producer is to be furnished with detailed recommendations and consultation on pest population management. Periodic refresher courses on new developments in pest control should be made available to extension personnel and professional consultants.

Some of the retraining necessary to develop the knowledge base for extension personnel in plant protection can be done through workshops, symposia, short courses, and the like. Certainly, the extension specialists in the separate disciplines will still be needed in even larger numbers than in the past. But also there appears to be a great need for field-level personnel trained much more broadly than in the past, who will be able to recognize the different plant disease and pest problems and at least call them to the attention of specialists in an intelligent fashion.

DIFFICULTIES OF FIELD DIAGNOSIS

Field diagnosis of plant disease and pest problems is exceedingly difficult. This is not usually appreciated by those who work most of their lives at desks or in laboratories. In particular, it's not appreciated by many administrators, nor by the general public who often do not understand the complexities of nature in the field. Nearly every experienced field man has had the distinct pleasure of traveling in the field with an often condescending, highly opinionated laboratory research specialist. It is revealing at such times to see the laboratory specialist flounder in his efforts at diagnosis in the field.

Because of these special diagnostic difficulties, there are many in plant pathology who think every student should be required to serve an apprenticeship working in a plant disease diagnostic laboratory and as a field extension plant pathologist. This is a very humbling experience and everyone will benefit from it no matter what his future plans may be. This is equally true in other areas in plant protection.

The difficulties involved in diagnosis are very well summarized in the following statement (Anon. 1972c):

> Positive and correct diagnosis of a disease is a prerequisite to the application of effective control measures. A well trained and experienced diagnostician should be able to make a systematic visual inspection of the plant disease situation and find the primary cause, or at least narrow it down to some of the most likely possibilities. Laboratory and greenhouse techniques may then be used to identify or verify the cause. Since the latter are time consuming and costly, the importance of thorough visual inspection is obvious. Far more can be learned and appreciated about a disease in the field than when working with dried specimens removed from their origin. Once in the field, one must have a thorough appreciation of the normal growth characteristics of the crop or commodity in question. Otherwise the "disease" situation may not be apparent. This is why diagnosticians prefer to work with a narrow range of commodities such as field, vegetable, fruit, or ornamental crops. Even then there is no substitute for experience. The accurate diagnosis of plant ailments is a skill that requires much time and effort to develop. It is particularly helpful to talk to the grower. Persistent questioning about the problem often uncovers some helpful clues to correct diagnosis and the pattern of distribution of the disease can provide valuable information.

Certainly, there is no substitute for the well-trained field man and he may well be the most difficult of all personnel to train adequately. It is relatively easy to train an individual in the use of a sophisticated piece of equipment in the laboratory, equipment that can be used under controlled conditions again and again. It is quite a different matter to teach an individual to assess accurately the myriad activities of uncontrolled "Mother Nature" in the field.

KINDS OF EXTENSION PERSONNEL NEEDED

What kind of extension specialists, then, are going to be necessary in the new coordinated and integrated plant protection programs of the future? E. Smith (1972) suggests that the greatest immediate need is for a new "extension pesticide use specialist," a person who is trained in the use of all pesticides (fungicides, insecticides, nematicides, and herbicides) at the field level. This individual could be working for an industry, as a private consultant, for a grower's cooperative, or for the government, depending upon the situation.

There definitely needs to be a hierarchy of personnel extending from the sophisticated research levels at the top down to the grower, in a meaningful and effective relationship. Sherf (1973) suggests for the United States that the state extension and regional specialists in all disciplines be continued but that they become more involved in applied or adaptive research activities in the field. He feels the need for laboratory diagnostic specialists in plant pathology as well as in the other disciplines. They would be practical diagnosticians who work primarily in the laboratory but are available to the field personnel for diagnosis of difficult field problems. These laboratories could be set up either at state or regional levels within states depending upon the complexity of the original situation and the amount of diagnostic work needed.

Another group of specialists should be involved only with extension training in the different disciplines and specializing in workshops, seminars, short courses, and the like. These would train pesticide specialists, other field specialists, and growers. The continuing training of generalists to function at the field level probably should be done by groups of specialists working specifically in practical programs required for the general training of field-oriented personnel.

At the field level, Good (1974a) feels that pest management education and services can be delivered by one or more of the following type programs:

1) Public sector pest management programs operated by the cooperative extension service involving state and county staffs and assisted by regulatory agencies such as the state department of agriculture.

2) Grower-operated and controlled associations, pest management districts or other organizations of the cooperative type.

3) Employees of processors or corporate farms.

4) Individual consultants or private consulting firms.

5) Technical representatives of the agricultural chemistry industry including retail pesticide salesmen who have knowledge of, and responsibly can advise on, integrated pest management.

In addition, many feel that educational personnel trained to utilize the newspaper, radio, TV, and other public media are very badly needed not only for public relations activity beamed at the urban populations but also to inform the farmers *and* the general public (urban gardeners) concerning the practical problems of plant protection. Such personnel could be added as consultants for educational purposes to city or state government staffs or could be hired by large urban TV or radio corporations.

EXTENSION REORGANIZATION—PUBLIC AND PRIVATE

One thing is certain, there must be an efficient "pipeline" system that functions rapidly from the top down with meaningful, well-organized, practical plant protection information. There must also be another "pipeline" that functions with equal efficiency, operating from the bottom up, which carries problems and questions concerning field problems from the grower up to appropriate levels for quick solutions to tough problems. Any system which fails to funnel practical research and field information rapidly to the grower, and which also fails to solve his problems quickly at the field level, is not going to be acceptable.

Sherf (1973) feels that the problems of extension are becoming so complex that it is not possible for the typical county agricultural extension system to function in a modern technological society. He envisions two types of extension services: one aimed primarily at commercial agricultural problems and one aimed at the "people programs" and the suburban environment.

The agricultural extension services are being reorganized in most states in the United States, altering the pattern of the past as it relates to plant protection and also in many other aspects of agricultural extension. There are many examples of new organization structures being attempted. One that concerns plant protection has been initiated in Kansas (Knutson 1975). In Kansas, extension specialists have been assigned to serve regions. These regions are made up of several counties of the state. The plant protection specialist is trained or is being trained to serve all needs as a generalist and to handle insect, plant disease, nematode and weed problems. As in the past, the extension specialists in the various disciplines are being maintained at the land-grant university, and some specialists are now attached to specific regions of the state based upon need. The county extension programs working within the regions are functioning as they have in the past but with the additional help of regional plant protection specialists.

At the field level, scouts are hired in the summer to work in assigned areas on specific crops according to need. These are either college science students or school teachers, often biology teachers from local high schools. They are given short-course training on specific plant protection problems relevant to the areas where they will work. This program was initiated in 1974. The first year's effort was so successful that all involved wished to see it made a permanent feature.

A mobile Plant Health Diagnostic laboratory is an interesting extension development in Oklahoma. A large four-wheeled trailer laboratory is used. Growers get better and quicker service, and graduate students have a fine facility in which to pursue their field internship. Also, the university's public relations image has been greatly improved by this technique (Anon. 1974).

In all parts of the United States pesticide salesmen and field representatives of pesticide companies are valuable sources of information to the grower and undoubtedly will continue to be. However, as mentioned earlier in this discussion, there is also a new private consultant role developing in crop protection and pest management, particularly in entomology where the field possibilities have developed much more rapidly than in other disciplines. This private practitioner or consultant may work as a general practitioner but, more likely, he functions as a general pesticide specialist. More of these individuals are functioning as general consultants for crop protection and pest management each year. As pointed out by Good (1974c) the larger growers or producers of the high-value per acre crops are much more likely to seek out the private consultants than are small growers or those growing low-value per acre crops.

Sherf (1973) documents the development of the private consultant in the field of plant pathology and discusses his probable future in agriculture.

> In the 1970's a new role is developing for extension plant pathologists and may become of considerable importance in the near future, one that may eventually mean the partial transfer of public plant pathology to the private sector. With the furor over the use of pesticides for control of pests in both agricultural and urban-suburban situations, several states have passed legislation requiring that certain pesticides be available and sold only on written prescription by licensed professionals. The task of training, testing and licensing this corps of personnel will fall largely on extension, and the next step will be for former extension plant pathologists themselves to enter this new service field for a fee somewhat in the manner of the professional veterinarians. This would take our profession out of the realm of "socialized plant medicine" (where it has always been until now) and into the private arena as with human medical services in the USA.
>
> Another factor giving impetus to such a move is the need for frequent and time consuming services by large farm cooperatives who must have precise recommendations on disease control in their farming operations and who are very willing to pay well for such private service. This, too, may be a profitable profession in the near future. So service plant pathologists in the 1980's may well be of three types: namely, private diagnosticians available on a fee basis for urban, suburban, and rural citizens, private salaried pathologists working for farm cooperatives, large farm commodity groups and so forth; and traditional extension pathologists paid by federal and state funds. This latter type of extension worker will probably be supported partially by farm and agri-business grants.

The efforts to revise and alter agriculture extension service activities in plant protection in the United States for the most part are applicable to other parts of the developed world. Each country and area varies the emphasis somewhat but certainly the efforts in the United States are illustrative of efforts going on in most other developed countries.

EXTENSION SERVICES IN THE DEVELOPING WORLD

In the developing world, however, the role of agricultural extension varies tremendously. In the more sophisticated small countries such as Israel and Taiwan (The Republic of China), discussed earlier, the agricultural extension services that have evolved and which serve the small growers of these countries are very different, but in their own way they are as sophisticated as anything devised in any of the developed countries.

By contrast, in other developing countries there are many cases in which not a single trained agricultural extension worker exists. Schultz (1974) reported that in Southeast Asia there was only one trained extension plant pathologist outside of Taiwan and Japan. Although this might be questioned, it is certainly true that there are very few trained extension people in plant protection in Southeast Asia, and many of those attempting to function as such are trained inadequately. As Schultz points out, there is a lack of local information on plant protection and there is very little applied field research that is practical and aimed at solving immediate food production problems. There is tremendous need for applied, adaptive research aimed at the development of simple, socially and economically acceptable plant disease and pest control methods. The tragedy is that there is very little inclination on the part of most scientific personnel in these countries to become expert in such research at the field level. In most instances there is a complete lack of accurate information on the prevalence of particular diseases and pests, and the amount of losses they cause. This type of information is absolutely essential for intelligent analysis of field problems.

The post-harvest diseases and pests in the hot and humid

tropics are particularly troublesome. Typically, there is a lack of refrigeration, storage facilities are primitive, transportation is slow. All of these aggravate the problems of post-harvest destruction (Schultz 1974).

Several recent developments in agriculture in various parts of the developing world have made plant diseases and pest problems more severe. First there is the large-scale production of the high-yielding varieties that require pesticides and fertilizers to achieve their full yield potential. The trend toward the use of these varieties and the consequent movement toward larger, extensive monocultures containing much more uniform germ plasm, also increases the threat of pests and plant diseases. The second trend is toward more intensive utilization of land, changes in land tenure which are moving toward fewer, larger holdings with, in many instances, less efficient utilization of land. Third is the trend toward multiple cropping (Schultz 1974).

There is a growing realization in the developing world that some pesticides are hazardous to man and the environment. Some countries are developing pest control legislation and regulations even before they have a trained extension staff to apply the pesticides or to regulate their use. Tremendous educational efforts will be required not only to train specialists to work in the field but also to train growers. These educational efforts will be needed before regulations associated with the use and/or abuse of pesticides can be enforced realistically.

All of these problems mean a far greater need for trained extension personnel in the developing world in the future. When one considers the extreme shortage of such personnel now, the problems that loom ahead are truly staggering. When one also contemplates the impending food shortages in portions of the developing world and the burgeoning populations, the problems become even more overwhelming.

Sherf (1973) discusses this problem:

> The attempt to carry large quantities of food to the world's hungry has been essentially a failure. The logistics and the cost of transport alone are overwhelming. By contrast, the self-help

> programs sponsored by the Food and Agriculture Organization of the United Nations, the Rockefeller Foundation, the Ford Foundation and others that are developing and teach new technology and extension methods are finally yielding results.
>
> The talents of the extension plant disease specialist are sorely needed in these food deficient countries. Plant pathologists skilled in applied research and especially in extension will find attractive opportunities to apply their talents to the compelling needs of the developing countries. This will require men with a true missionary spirit since monetary and professional recognition most likely will be meager.

Unfortunately, most of the students studying agriculture at graduate schools in the United States who come from developing countries are not being trained in applied, adaptive research techniques. Most of them do not wish to cultivate this area of training. The vast majority intend to become sophisticated laboratory scientists in narrow scientific specialties. This training often is nearly useless when they return home. The great need of most of their countries is for practical, applied, adaptive research. These needs are largely ignored by highly specialized individuals who are not field-oriented or trained. In fact, these people are often antagonistic toward any field activities. Consequently, one of the greatest needs is to stimulate, or even force, the movement of large numbers of students into extension-type training programs that include an adequate preparation for adaptive, applied research techniques. There also is the necessity of convincing the administrations of most of the developing countries that applied, adaptive research in agriculture is absolutely essential and that its neglect will result in catastrophe.

For countries characterized by small land holdings, The Republic of China (Taiwan) has developed one of the most outstanding agricultural extension systems. This is described by Lo (1969):

> Plant protection technicians at various levels of government and farmers' associations play an active role in plant protection extension. They disseminate the scientific knowledge developed by researchers to the farmers and bring back field problems to research workers. In order to maintain the efficiency of the

overall program of plant protection, technicians should receive on-the-job training periodically so as to keep their knowledge up-to-date.

The small farm agriculture of Taiwan puts the island in a situation similar to that in many other Asian countries, for example, large numbers of farmers versus small size of land holdings. Since it is difficult to teach farmers modern knowledge individually, cooperative pest control becomes necessary. In 1960 cooperative pest control teams for rice were first organized, each covering 100 hectares of paddy field, with operations under the strict supervision of the government. It was found that cooperative control could result in higher rice yields, and also in lower cost because pesticides of standard quality could be purchased at a bargain. In view of the success achieved with rice, cooperative control teams for citrus and vegetables have also been established and the advantages of this practice have been generally recognized. It is proposed that the government should make further efforts to strengthen the system for more fruitful results in plant protection.

It is interesting to note that in The Republic of China (Taiwan) most of the plant protection technicians are at the bachelor's degree level or below. They are given largely onthe-job training through short courses. They function at the field level very close to the farmer and to the farmer cooperative groups. They help analyze the problems they see in the field, or they quickly call in specialists when their knowledge is not sufficient to assess the local field problems (Lo 1969). It is important to emphasize that most countries in the developing world simply do not have this trained person working at the field level who is closely associated with farmers or cooperative farmer groups.

Apple (1971) has suggested, as have several others, the development of an international extension and/or adaptive research network for pest management. This is an excellent idea, particularly if it emphasizes applied, adaptive research, and a solution of real problems occurring at the field level. He suggests that,

Land-grant universities can be the centers of such networks in developed countries and the international research centers in the developing countries. The main need is to develop a coordinated system of pest management activity. The network would pro-

> vide the means for training domestic personnel at all levels required for the country to protect its crops against pests, the health of the people against unwise use of pesticides and the quality of the environment through the utilization of appropriate pest control methodologies. Participation by developing countries in the research network would be a means of securing additional protection for the production gains of the "green revolution." An important function of the network would be to coordinate project activities with related programs of other agencies, foundations and governments. One objective would be to have the pest management research network serve as a communications medium between the many independent program activities currently active in the developing nations. Cooperation or liaison would be established with the international research centers (for example, CIAT, CIMMYT, IRRI, IITA, and others) international development agencies, (for example World Bank, FAO, UNDP, USAID, and so forth) private foundations, bilateral country programs and in-country research programs with regional or international significance. Coordination would be enhanced by the formation of an informal "inter-pest management group."

This is a fine idea and should be implemented, but it should be combined somehow with present extension programs since agricultural extension also is primarily concerned with applied, adaptive research occurring at the local level. What is needed is an International Center for the coordination of applied, adaptive agricultural research information. A good suggestion would be to have an information and coordination center within an organization such as FAO, United Nations, Rome. Out of such a center the agricultural information could go to all parts of the world through satellites, radio, telephone, television, even via "ham" radio operators, plus the usual brochures and bulletins. This world coordination and communication center for applied research and extension information in agriculture (including plant protection) could perhaps be completely computerized and could function as the top of the information "pipeline" for a myriad of research and extension activities in the developing world. Utilizing all known electronic techniques, it would eventually be possible to reach via transitor radio the most primitive villages in the developing world with accurate oral information useful at the local level. It would also be possible to furnish information to extension personnel at all levels utilizing an electronic instructional system operating from the FAO center or from various regional centers. This type of oral system is much more likely to be effective at the field level where most of the small farmers of the developing world cannot read or write. However, they are not stupid. They can listen to transistor radio programs that are presented accurately and simply in their own language, and through these programs they can gain tremendous amounts of useful information that can be applied at the field level.

In some countries of the developing world sophisticated centers of agricultural research and extension are being developed or will be developed soon. These should be staffed with local and international personnel in the area of plant protection. Schultz (1974) suggests the following improvements on the usual pattern of visiting scientists for the typical AID programs.

1) The assignment should be much longer. If possible, the extension man assigned to any foreign country should be on a permanent career basis but should be allowed to retain membership in departments at home. Also he should be assigned to reasonably well-equipped and staffed institutions with reasonable support for travel and the necessary extension activities.

2) The personnel selection should be done with much more care. Only those with a strong humanistic viewpoint and honestly interested in the particular country should be chosen. They should be people who are patient and understanding, highly adaptable to the diverse cultural backgrounds in the various countries and able to inspire others by their enthusiasm and competence. They should be able to direct their information successfully to the intellectual and experiential level of the native people.

3) They should have a very broad background of training and should be able to integrate various fields of agriculture at the field level in a practical approach to solve a problem. They should be given adequate training in the customs, history, language and culture of the country to which they

are assigned and should not be sent unless they are accepted by the local government and allowed to operate with only a minimum of political restriction.

Actually, the new plant protection centers being developed by FAO (UNDP) United Nations, Rome, in cooperation with several Asian countries are centers very similar to those suggested by Schultz (1974). These could function very easily as regional training centers in plant protection, receiving a constant flow of information from a center in Rome via satellite or radio and relaying it verbally to the local levels in local languages.

The food problems of the world are great and certainly will become greater. It would seem that every effort should be made to develop an international system of applied, adaptive agricultural research and extension information aimed at the quick dissemination of usable information at the field level. Every possible modern technique should be utilized and completely international cooperative worldwide efforts should characterize the program.

Chapter 14

Some Concluding Thoughts

No amount of talk, laws, or fearsome "Doomsday" prophesies will make the pests and diseases of animals and plants disappear. Nor will the problems associated with them be solved without intensive effort on the part of all trained personnel working closely with the farmer. Environmentalists, most of whom are not well trained in either agriculture or science, often seem to have the "misty-eyed" notion that if we would just let nature alone we could live again in a world of pristine purity. Assuming that we could immediately destroy 80 to 90 percent of the human race, this is probably true, but that world would again be one in which the average lifespan would be fewer than 30 years and the miseries of crop losses and many recently conquered human diseases would again become commonplace. In short, the "good old days" were not so good, and it is not likely that even the most avid environmentalists would voluntarily go back to them. The reality is that we must live with pests, diseases and the like and where needed do anything possible to control them.

McNew (1972) summarizes this matter well insofar as the future need for plant protection is concerned:

> The harsh reality of the situation is that we must live with pests be they insects, mites, snails, worms, fungi, bacteria, viruses, epiphytic plants, allergens or weeds. Rarely do we eradicate them. The best we can do is to co-exist with them. We and they are part of a giant ecosystem, and man is simply trying to shift the balance of power so he can exist in reasonable comfort and security. It is necessary for us to be sufficiently objective to understand this ecosystem in which the role of man is defined as a competitive force, albeit a bit more intellectual and far-sighted than his competitors. Perhaps it would be better to remember that it is our strategy for survival in competition with biological forces that we cannot crush or obliterate. We merely try to outwit their natural propensity for multiplication and to constantly readjust our strategy to their adjustments.

Man, it is clear, is in fierce competition for food with other living organisms on earth. Unfortunately, food is often in short supply, for man as well as for other species. Certainly, the move toward population control of man is sensible and necessary in the long run, but even with lowered birth rates the battle for food against other organisms will continue. For example, man has been battling the insects for centuries. If there is an endangered species among this group of enemy insects, it has not been brought to our attention. The best that we have been able to achieve thus far is the suppression of populations in a given location. And this is probably the best we can hope to do. In the developing world well over half of the food produced is lost to pests or plant diseases, either in the field, in storage, or in transit. In the developed world this loss has been cut to approximately 25 percent. This is still a discouragingly high figure (McNew 1972).

With these facts in mind, McNew (1972) makes the following suggestions:

> The basic assumptions that seem valid in planning our strategy within which we should plan our tradeoffs, are as follows:
>
> 1) Modern intensive agriculture is here to stay and as long as fuel is available, it will be mechanized. Therefore, monoculture of crops, large livestock feeding operations, and other intensive developments are a necessity if mankind is to be freed for maintaining and expanding his industrial empire and practicing the creative arts.

2) Soils are to be managed principally to stimulate productivity and only secondarily as a pest control operation but they can be modified to avoid hazardous conditions.

3) Crop and livestock genetics will have to be manipulated but with the first objective of avoiding pest inducement.

4) Every weakness in the pest's biocycle must be exploited to suppress its survival, reproduction, and multiplication.

5) The environment must be managed so as to suppress or destroy the primary pest or to alleviate its attack by altering ecological conditions where possible or by using properly designed chemicals when necessary.

The cost of plant protection can be so high in certain cases that it becomes prohibitive. Sometimes research people working in the disciplines tend to ignore or forget costs—always to their sorrow. This point is emphasized by Robins (1972):

> The first national (also international) priority is to place at the disposal of the consuming public an adequate supply of safe, wholesome and nutritious food at a price commensurate with its ability to pay. Second in importance is to advocate an economically viable agriculture. Without economic viability it cannot be effective. Finally, our agriculture must be sound, sound technically, structurally, economically, and environmentally. Unless this is so it cannot be effective or economically viable. Any crop protection or pest management program whether it be developed by individual disciplines or cooperative on an interdisciplinary basis must face up to the reality of economics in agriculture and the control measures that are devised must fit into the apparent economic constraints. If our control procedures as devised are too expensive, then we have no alternative but to advocate the kind of compromises that are necessary to achieve the best possible control. We must make the hard choices to seek a balance between what is best, what is most economic and what is easiest.

Actually, the cost problem is one of the most important reasons for emphasizing the development of simpler and cheaper interdisciplinary, combined or integrated control procedures for pests and diseases. This fact should be remembered constantly by those working in the various disciplines of plant protection.

Recently, there has been a call for an end to the use of some pesticidal chemicals and other biocides, largely by environmental groups. Certainly, a few of the most dangerous chemicals can probably be eliminated gradually, but there is no hope of eliminating the use of all pesticidal chemicals in the foreseeable future if we are to maintain our present levels of food production or increase them to feed a rapidly enlarging human population. This necessity for the continued use of chemicals is stressed by McNew (1972):

> The use of chemicals to disinfect, to suppress, to destroy the pest or to permit the crop to escape the most extreme ravages will remain a potent force. Without this form of insurance most agricultural practitioners cannot afford to invest in expensive equipment, precise fertilizer components and tailor made crop varieties. All of these things can be afforded only when the farmer takes out crop insurance of several sorts, including the suppression of pests. These chemicals of the future are going to be more expensive because they must be more specific so as to avoid adverse side effects, must meet rigid standards of environmental safety, and must be distributed and used with better control than at present. The general pesticide will probably give way to the more specific types as demands for safety gain prominence.

R. Smith (1972) also stresses the necessity for continued use of pesticides.

> In my stressing of manifold control techniques in plant protection systems for agriculture in the tropics and subtropics (as well as elsewhere), I wish to make it very clear that this does not mean the substitution of biological controls for all chemical controls. It is true that I strongly advocate a greatly increased interest in and support for biological controls and other neglected controls appropriate to the better management of pest populations, but this does not even infer the elimination of chemical controls. That, of course, would be unrealistic because it does not take into account the facts of the real world.

In the developing world where it has been and continues to be very difficult to procure pesticides even when the money is available to buy them, the attitude toward pesticidal chemicals has been different and probably will remain so. In these areas there has been little or no movement away from the use of pesticides nor, for the most part, any effort by political or pressure groups to have them curtailed. Buddenhagen (1975) discusses this difference:

> A major basic difference revolves around the probable fact that increased production in the tropics will require increased use of pesticides and fertilizers rather than decreased use, whereas such further increases in the United States and Japan are probably not desirable objectives or even main objectives.

Pesticides are not available for most of the small farmers of the developing world. However, in areas where they are available, their use typically has provided dramatic increases in yield. This is emphasized by Brady (1974):

> Partly controlling the rice stem borer in the Philippines by spraying with insecticides produced minimum yield increases of about half a ton per acre, doubling the average yield in Asia, according to a study by the International Rice Research Institute. But, because of their high cost, relative scarcity and difficulty of application, herbicides and insecticides are considered at best only a partial solution to the problem of field pests in the developing countries.

Despite the fact that pesticides often will be used as absolute necessities in the developed world and will be used more and more, where possible, in the developing world, the long-term efforts of plant protection workers should be toward developing techniques which gradually eliminate or minimize their use. There are many legitimate reasons for this gradual movement away from pesticides, among them is the destruction of valuable and useful insects and other arthropods, the dangers to farm animals, the development of highly resistant insect and mite species after prolonged exposure to some chemicals, the consequent necessity of using more and more applications of pesticides per season, and finally the great and increasing cost of pesticides to the grower and the consequent desire of every grower to minimize or eliminate their use. Consequently, this will undoubtedly be a long-term goal of major significance.

There are also a few legitimate cases of potentially dangerous accumulations of pesticides in nature. These situations must be carefully studied, analyzed, monitored, and eventually eliminated. For all of the above reasons there will be a strong effort to develop techniques for safer handling of present pesticides, to develop new pesticides which minimize or eliminate the present hazards and, eventually, to move away from the use of pesticides whenever possible.

A clear necessity is to move quickly toward a coordinated, interdisciplinary, integrative approach to plant protection which includes every possible combination of control procedures for any given plant disease or pest. Specifically, the search must be for the cheapest and safest possible methods of control and, hopefully, for control procedures that are not only inexpensive but essentially automatic, such as those achieved through routine cultural practices, biological control forces, or those brought about through the breeding of resistant or immune plants.

If pesticides or other materials are to be applied, the ideal procedure would be a single application that could function as a systemic poison against a specific plant disease organism or pest, or one that would have enough residual potency to be released slowly at a satisfactory level for adequate control of the disease or pest. These goals require a new interdisciplinary and cooperative approach toward the total problems of plant protection in each crop, an approach which, unfortunately, has been rather rare in the past.

The various disciplines associated with plant protection are clearly combined at the field level. The grower, unlike the research specialist, must face nature as a whole, and any approach that will make it simpler and cheaper for him to achieve the necessary control or suppression of plant diseases and pests will be a clear-cut advantage in an always tough battle. These control or management techniques will be more desirable to the grower if they combine controls for as many diseases and pests as possible in a single procedure. The soaring costs of modern agriculture will make growers anxious to move toward any technique that will cut costs and yet achieve adequate management or suppression of crop diseases and pests.

Many feel that these pressures for better, cheaper, simpler techniques emphasize the necessity for the training of a new professional, a general practitioner in plant protection. This general practitioner would function as a consultant to growers. He would be serviced in a variety of ways by specialists now functioning effectively in the various disciplines. This individual would fulfill the approximate role of the general practitioner in human medicine and would attempt to be an objective "broker" serving the interests of the farmer and the grower.

Other pressures toward creating a new general practitioner are the new and more restrictive laws associated with pesticide uses, and the clear movement toward certification of pesticide applicators. Some of the restricted pesticides in the foreseeable future probably will be available only on a prescription basis, or they will be applied by licensed pesticide applicators under the supervision of highly trained plant protection specialists.

Because of this apparent need for a general practitioner, many educational institutions in the United States and elsewhere recently have developed or are developing bachelor's and master's degree training programs to produce such specialists. It is not clear yet just how rapidly these trained people can procure satisfactory positions. In those situations particularly associated with severe entomological problems, there are now quite a few practitioners finding work and some are entering very renumerative consulting practices. The other disciplines associated with plant protection are moving more cautiously in the direction of the general practitioner since it is not clear as yet exactly how these individuals will find employment in the field.

In recent years there has been a slow but steady movement toward the reorganization of research, teaching and extension in plant protection disciplines toward interdisciplinary and cooperative solutions to severe problems. As indicated above, new teaching programs for the general practitioner are evolving in some institutions. Also, there has been the recent development of a good number of complex, highly sophisticated interdisciplinary cooperative research projects, usually built around the needs of a specific crop. In particular, the large international foundation-sponsored agricultural research organizations throughout the developing world have shown the way in organizing their research for crop improvement, production, and management around a single crop and its specific problems, and they have developed highly effective research teams for the solution not only of plant protection problems but all problems associated with growing a high-yield, well-adapted, high-quality crop.

In United States agricultural extension programs there has been a recent movement toward reorganization of plant protection services. In some states there are now regional personnel rather than state specialists. This is intended to bring the advisory services closer to the farmer and thereby make them more effective. On some crops such as cotton, apples, peaches, potatoes, and a few others, rather sophisticated field survey and monitoring programs have been designed that are essentially interdisciplinary, cooperative ventures in developing more realistic advisory services centered at the field or grower level. The movement in this direction will undoubtedly continue on crops where such efforts prove to be possible and profitable. Here again, the economic constraints of modern agriculture demand more efficient, cheaper techniques and more realistic advisory services that respond more rapidly to the grower's immediate needs in a specific area in a given season and on a specific crop.

Some of the more sophisticated research and extension programs on important crops, in which the crop value is high, have moved and are moving toward a systems approach to all agricultural efforts in the field, including plant protection. Here the most sophisticated computer methods of systems analysis and simulation modeling are being developed and in a few cases have proved to be of practical use. Without question, this movement will expand in the developed countries that are characterized by large farms and large acreages of specific crops of high value per acre.

Even though this movement toward interdisciplinary cooperation and the utilization of combination or integrated control procedures will undoubtedly continue and intensify, this does not mean that the separate disciplines now supporting plant protection will disappear. Actually the basic research that is needed in each of these disciplines to support the applied research at the field level needs to be supported much more than at present, and the advice from specialists in these disciplines will be needed even more by the general practitioners functioning at the field level. The development of the general practitioner will spur additional development in teaching, research, and extension in each of the supporting disciplines simply because of the amazing complexity of the field problems encountered.

Disease organisms and pests, particularly the microorganisms, are tough and tricky enemies. Most of them can and do evolve much more rapidly under adverse pressure than larger and more complex species. Consequently, it will be necessary to continue every effort to develop new control and management techniques as older techniques prove to be inadequate because of changing biotypes. The fear, often expressed, that the development of a new, more generalized discipline will mean the demise of others is completely groundless. Actually, the development of a new general field will stimulate the need for increased efforts in the supporting specialties just as it has in human medicine.

The problems associated with plant protection seem complex enough in the developed countries in the temperate zones but, in reality, they are far more complex in the developing world, particularly in the tropics. As Wellman (1968) reported, there are many more publications and more scientists in the temperate zone, and yet far more crop species are grown in the tropics and there are many more diseases and pests. On tomato, for example, he lists 32 diseases of importance in the temperate zone, whereas there are from 50 to 278 in various areas of the tropics. On cabbage, he lists nine diseases of economic significance in Wisconsin, and 18 to 36 diseases in various areas of the American tropics. On citrus trees, 50 diseases of economic consequence have been reported and studied in the United States and other subtropical citrus growing areas, whereas in the West Indies and South America he reports 248 diseases of economic significance.

Also, the plant protection problems of the developing world of the tropics are generally far more complex and usually require different techniques for control because of great differences in environment, the absence of winter, and other variables not typical of the temperate zone.

Paddock (1967) discussed some of these unique problems of the developing world, particularly in plant pathology.

> New, more imaginative, disease control methods specifically designed for the needs of the developing world are needed. Plant pathology is essentially a western science. Its origins and growth have been largely confined to the temperate zone of the developed world. It is then almost to be expected that the agricultural programs of the developing world are often mirror images of what have proved successful in the developed world. The question asked here is whether or not plant pathologists have sufficiently studied alternative methods of disease control specifically adapted to the developing nations. A similar question might be asked of the other agricultural sciences as well. Efforts to date to find imaginative solutions for the agricultural problems of the hungry world have been largely by individuals. While many of these efforts have been remarkably effective, they have been too few and too sporadic to make a significant impact on the problem. A few symposia have been held, but the professional societies have not otherwise recognized the need for demonstrating leadership in this area. The American Phytopathological Society is a good case in point. It has had a long standing "International Cooperation Committee" chaired by some outstanding pathologists. The committee collected colored slides of tropical diseases, arranged to send subscriptions of the society's journal to two foreign experimental stations, and is currently preparing a glossary of Spanish-English terms in common use in plant pathology. Alas, none of this is of earth shaking importance.

R. Smith (1972) also feels that the scientists of the temperate zones have a clear and present responsibility to do their very best to help in the development of methods that are adaptable to the developing world of the tropics. He feels a moral responsibility is present to assist in the development of international plant protection systems that are appropriate to the local environments, the economies, and the social customs of the countries of the developing world. He feels strongly that these systems of management and control should be firmly based upon ecological principles and that combination and/or integrated control techniques should be devised. He emphasizes again that most of the farmers of the developing world are unable to procure pesticides even if they can afford them. The supplies just aren't there. Consequently, the cultural, biological, genetic, and other combination control possibilities need to be explored as much or more in the developing world as in the developed world where pesticides are routinely available.

The basic plant protection problems of the developing world are perhaps best summarized by Buddenhagan (1975):

> If the basic research and knowledge in ecology and epidemiology necessary to make "pest management" more rational seem to be inadequate in the temperate countries (advanced) this inadequacy is magnified a hundred to a thousand fold in most tropical countries. Yet, who in the tropics is responsible for seeking and funding such knowledge. In the face of such lack of knowledge, expenditures toward "pest management" or "crop protection" action programs will automatically result in increased spraying of pesticides, in spite of the known desirability of the pure theoretical concept of "pest management," while at the same time lip service will be given to the concept. This has already happened in India.
>
> In most underdeveloped tropical countries, where funds and positions are very limited, the major concern of an administrator responsible for an increased emphasis on crop protection and pest management will be in maintaining or enlarging his organization and its budget in tight competition with many other agencies. Seen as support in this struggle are increased numbers of acres sprayed, helicopters bought, and increased state-supported purchase of pesticides. If crop yields go up, such activities can take partial credit, whether true or not. The slow accretion of important biological-ecological, epidemiological information will not sell easily.

The foregoing quotations outline the problems of the developing tropical countries and make clear the fact that they are actually greater problems than those faced by the developed countries in the temperate zone. Certainly, these problems will not just go away. Instead they will become more intense as food needs increase. An entirely new and far more imaginative interdisciplinary approach is required to solve the amazingly complex plant protection problems of the developing world. Compounding these problems will be the ever present fact of poverty and the consequent unavailability of important inputs needed to control crop pests and diseases. Further compounding the problems will be the low level of literacy among most of the rural people, the generally poor state of health, the difficulty of transportation, poor communications, and many other factors.

A new international approach is needed, and the recent effort by the FAO (UNDP), United Nations, Rome to develop plant protection centers in several Asiatic countries is a strong step in the right direction. Also, recent developments in enlarging foundation-supported international agricultural research and extension programs on important tropical crops is highly significant.

However, an even larger international program is needed. Perhaps this program could be built around an international clearing house or center for research and extension information in crop improvement, production, and management, including plant protection. Using every known electronic media, including satellites, it might be possible to develop a system of rapid verbal communication that could move research information rapidly down from the top to the grower level even in remote areas via strategically placed transistor radios. Radio communication on radio telephone could also move local problems up to regional, state, national, or international levels quickly to appropriate specialists for consideration and, hopefully, quick solution of severe local or regional problems. The development of such an international clearing house and communication center for coordinating practical scientific information related to agriculture should be one of the first items on any international agricultural agenda.

It is clear that the system devised must be an efficient *verbal* system of information transfer. The written or brochure systems of the past are important and should be continued, but they are not adequate to solve the problems of the illiterate millions of small farmers of the developing tropics. A verbal system must be devised utilizing all known electronic media techniques to get needed information quickly to the farmers and to move their field problems quickly to the appropriate level for consideration and rapid solution.

Finally, the many disagreements among the disciplines representing various facets of plant protection would disappear rapidly if each were willing to admit that plant protection is only a part, and a relatively small part, of the total agricultural input for growing a healthy crop. The best research and extension organizations now are organized in a way that recognizes this. The large and extremely effective international foundation-supported programs, the largest and best plantation organizations and industries, and many of the most effective land-grant institutions have designed their programs around the problems of a given important crop. Research teams have been hired to function cooperatively on the solution of problems associated with raising that crop. The primary stress has not been just plant protection but crop improvement, production, and management. All other subservient questions, including plant protection, have been dealt with in the overall effort to solve the many complex problems associated with growing a crop of the highest possible yield and quality.

This important position has been expressed by various workers in the field. For example, Hooker (1975) puts it this way:

> The plant science research worker in state experiment stations and federal laboratories must not lose contact with the crop and the way farmers grow it. He simply cannot become wholly academic and isolate himself from agriculture. Continued emphasis is needed on research with application to production, so that yields can be improved and stabilized. Unfortunately, this type of research is not held in high esteem by some scientists, administrators and even study panel experts, but with greater food needs, foreign trade and higher prices attitudes are changing.

Nusbaum and Ferris (1973) emphasize a similar theme:

> Because the cropping system is the dominant feature of agro-ecosystems, it is the foundation upon which integrated pest management systems rest. A well designed crop protection experiment may provide material for analytical study not only of nematode populations but also of other soil pathogens, soil insects and weeds. This emphasizes the interdisciplinary nature of pest management, especially where different kinds of pests interact with each other and with a given management practice. Hence, the concept of integrated control need not be limited to a combination of complimentary practices directed at a single topic but rather can be expanded to include pest complexes of varying dimension.
>
> Much progress is being made in breaking down the barriers that impede cooperation between disciplines and agencies. The emergence of broad concepts of pest management in the interest and involvement of all sectors, not only of research and educational institutions, but also of the agrobusiness community and the public at large, justify an optimistic outlook. The ecosystem approach to pest management presents great challenges and opportunities for the future. Models of biological systems are simplifications of the real world and are constructed to assist in understanding it. The application of systems analysis and simulation in predicting nematode behavior will provide a better understanding of their role in crop production. It will also aid in planning and evaluating integrated control programs fitting harmoniously into systems of land management designed to maintain the productive capability of the soil.

Another significant reminder on a related theme is presented by Janick, *et al.* (1969).

> The simple farmer of yesteryear is fast becoming a manager of crop production, drawing upon chemistry, biology, engineering, economics and many other disciplines. He runs a rural factory producing proteins, fats and carbohydrates. His goal is the same as that of any businessman, maximum return. Thus the farm operation of the future will be a high yield system, drawing upon many technical resources. All of the many activities may very well have to be integrated by business machine data processing programmed by specialists who never work in the field. This is already almost the case in advanced agriculture in the southwestern states (U.S.A.) Specialized services make such systems click. Much of American agriculture is already beyond

general servicing by county agents (farm advisors) trying to cover all facets with demonstrations, tours and bulletins. Experts in many fields are needed, operating as a team. Diverse inputs are demanded that must be correct, timely, sophisticated. The system is concerned with the whole sequence of steps dealing with the organism from growth to distribution and marketing. The agrobusiness manager must be aware of government programs and financial trends, as well as of newer varieties, cultural techniques and pest hazards. Eventually computers may relate these inputs and be used in all short and long term planning.

The foregoing statements are completely meaningful as they relate to the more technologically advanced, sophisticated agriculture programs characteristic of large farms in the temperate zone and some of the plantation industries of the tropics. However, in the developing world of the tropics what is needed even more are the tried and true methods of local, applied, adaptive research built around the very best statistically designed field plot techniques. By these standard research methods, carried out in every possible environment, it is and will be possible for the foreseeable future to furnish the necessary practical field information concerning crop improvement, increased production, and better management, including plant protection.

Buddenhagan (1975), after many years of experience in the tropics, emphasizes what he feels will be the greatest basic needs. Here again, the theme is built around the primary concern for crop improvement, improved production techniques, and better management principles.

In my view the most basic need, at least in food crops in the tropics, is to further develop the concept and practice that pest management is first and foremost the integration of the epidemiological pathologist and ecological entomologist and the crop production man with the practicing breeder who is breeding the next major food source for the pest. Towards this end I would put most of my money and most of my efforts at fostering coordinated research and development. For indeed, it is the breeder that now sets the stage of the play that will be played in the field by the pest and pathogens. Secondly, I would back epidemiological and ecological crop production oriented re-

> search areas, now extremely deficient in the tropics. Thirdly, I would attempt to develop agricultural education in these two directions and to so educate the very best young men thoroughly, wherever in the world it may be possible, with these emphases so that they can return and carry the zeal for such an approach to crop improvement in the tropics.

Buddenhagan continues with an extremely significant statement concerning plant protection that would be adaptable both to the developed and the developing world of the tropics, but particularly important in the latter.

> I believe it would be useful to ask ourselves what we are really trying to do under the term "pest management." Let us back up a moment. Presumably crops have a certain *yield potential.* On this criterion, new varieties are touted. Such varieties, when grown by a farmer, have a certain *actual yield.* The difference between *yield potential* and *actual yield* is where we want our pest management people to be concerned to do something immediate. This difference may be very large or very small and it may be due mostly to biological populations (of insects, pathogens, or weeds) or it may largely be due to a physiological disease (soil nutrient imbalance, inadequate soil fertilization, inadequate water management or bad weather conditions or other factors) normally of concern to the agronomist and soils man. No "pest management" practitioner will be able to sleuth such complexities in the tropics. In this view, it can be seen that we are really concerned then with overall *crop production* practices and especially for food crops with *crop improvement* (breeding new varieties). The real need then is to sleuth the intricacies responsible for the differences between *yield potential* and *actual yield* naturally without doing anything else. I believe that "pest management" as generally discussed cannot cover this real and complex need. I believe that to develop "pest management" only as a practitioneering operation for "the management of pest populations" as a separate entity from *crop production* and *crop improvement* and from *basic ecological research* is to depart from the most basic need of agricultural improvement in the tropics.

Certainly, those who have struggled for years to help improve the crops and agricultural practices of the developing tropical countries would applaud Buddenhagen's statement. Without question, the research orientations in plant protection of the future should be aimed in the direction Buddenhagan outlines: if possible, toward completely automatic methods of control that are built into the crop or cropping practices.

References Cited

Adkisson, Perry. 1974. *Opportunities for professional entomologists in private practice.* Bull. Entomological Soc. Amer. 20(4): 277–78.

Anonymous. 1965. *Restoring the quality of our environment.* Report of the Environmental Pollution Panel. President's Science Advisory Committee. Washington, D.C.: The White House.

———. 1966. *Proceedings of the FAO symposium on integrated pest control, III.* (October 1965). Rome: FAO, United Nations. 129 pp.

———. 1968. *Principles of plant and animal pest control. Vol. I: Plant disease development and control.* Publ. 1596. Div. of Biology and Agric. Washington, D.C.: National Academy of Sciences–National Research Council. 205 pp.

———. 1969. *Report of committee on persistent pesticides.* Div. of Biology and Agriculture. Washington, D.C.: National Academy of Sciences–National Research Council.

———. 1971. *Report of the 3rd session of the FAO panel of experts on integrated pest control.* Meeting Rept. AGP: 1970/M/7 (Sept. 10, 1970). Rome: FAO, United Nations.

———. 1972a. *Implementing practical pest management strategies.* Proc. National Ext. Insect Pest Management Workshop (March 14–16, 1972). College of Agriculture, Extension Service. Purdue Univ., Lafayette, Ind. 206 pp. (Sponsored by Extension Service, U.S. Dept. of Agriculture, cooperating with State Extension Services.)

———. 1972b. *Integrated pest management.* Stock No. 4111-0010. Prepared by The Council on Environmental Quality. Washington, D.C.: Supt. of Documents, G.P.O. 41 pp.

———. 1972c. *Study guide for agricultural pest control advisors on plant diseases.* University of California, Div. of Agricultural Sciences, Berkeley, Calif. 232 pp.

———. 1972d. *Systems of pest management and plant protection.* RICOP Committee on Plant Protection, St. Louis, Mo. (June 1972). A report from the Workshop for the Development of the Educational Concepts for an Integrated Approach to Economically and Ecologically Sound Pest Management and Plant Protection. South Miami, Fla.: BM&M Associates.

———. 1973. *Report of the 4th session of the FAO panel of experts on integrated pest control.* Meeting Rept. AGP: 1973/M/5 (Dec. 6–9, 1972). Rome: FAO, United Nations.

———. 1974. *Diagnosis on wheels.* Phytopathology News 8(11): 1–3.

———. 1975a. Untitled. Phytopathology News 9(12): 7.

———. 1975b. *Plant studies in the People's Republic of China.* A trip report of the American Plant Studies Delegation. Washington, D.C.: National Academy of Sciences. 206 pp.

———. 1975c. Untitled. Phytopathology News 9(9): 6.

———. 1975d. Untitled. Phytopathology News 9(12): 10.

———. 1975e. Untitled. International Newsletter on Plant Pathology 5(2): 1–8.

———. 1975f. Untitled. Phytopathology News 9(10): 9.

———. 1975g. *Pest control: an assessment of present alternative technologies. Vol. I: Contemporary pest control practices and prospects.* Environmental Studies Board. Washington, D.C.: National Academy of Sciences–National Research Council. 506 pp.

——, 1975h. *Plant health technology course.* Phytopathology News 9(5): 1, 7.

——. 1975i. *Plant protection.* Phytopathology News 9(7): 24.

——. 1976. *APS to take action at Kansas City on forming the ISCPP.* Phytopathology News 10(6): 2.

Appa Rao, A. 1974. Personal communication. Research Director, Andhra Pradesh Agricultural Univ., Hyderabad, A.P., India.

Apple, J. L. 1971. *A network approach to collaborative pest management research involving U.S. universities and developing nations.* Conference organized by Southeast Asia Development Advisory Group (SEADAG) (Nov. 1-3, 1971). Rural Development Panel, Bangkok, Thailand. Raleigh, N.C.: College of Agriculture, North Carolina State Univ.

——. 1972. *Intensified pest management needs of developing nations.* BioScience 22: 461-64.

——. 1974. *Integrated pest management: the status of research and academic programs.* Paper presented at the Agricultural Research Institute, Annual Meeting, Denver, Colo. (Oct. 15, 1974). Raleigh, N.C.: College of Agriculture, North Carolina State Univ.

Arneson, P. A. 1975. Personal communication. Dept. of Plant Pathology, Cornell Univ., Ithaca, N.Y.

Bennett, C. W. 1973. *A consideration of some of the factors important in the growth of the science of plant pathology.* Ann. Rev. Phytopathol. 10: 1-10.

Blasingame, Donald. 1974. Personal communication. Mississippi State Univ., State College, Miss.

Borlaug, Norman. 1972. Personal communication. International Maize and Wheat Improvement Center (CIMMYT), 8 Londres 40, Mexico 6, D.F., Mexico.

Boyer, W. P., L. O. Warren, and Charles Lincoln. 1962. *Cotton insect scouting in Arkansas.* Ark. Agri. Exp. Sta. Bull. 656. Univ. of Arkansas, Fayetteville, Ark.

Brady, Jane E. 1974. *Experts for pest control to increase world's food.* The New York Times (Oct. 28, 1974).

Brann, James L., Jr., and James P. Tette. 1973. *New York State apple pest management project, annual report for 1973.* New York Cooperative Extension and New York Agric. Expt. Stations in cooperation with A.R.S., U.S. Dept of Agriculture, Geneva Experiment Station, Geneva, N.Y. Unpublished. 37 pp.

Bruehl, George W. 1976. Personal communication. Dept. of Plant Pathology, Washington State Univ., Pullman, Wash.

Buddenhagan, Ivan C. 1975. Personal communication. Dept. of Plant Pathology, Univ. of Hawaii, Honolulu. (New address: IITA, Ibadan, Nigeria.)

Bunting, A. H. 1972. *Ecology of agriculture in the world of today and tomorrow* in *Pest control, strategies for the future.* Div. of Biology and Agriculture. Washington, D.C.: National Academy of Sciences-National Research Council, pp. 18-35.

Campbell, Robert W. 1972. *The conceptual organization of research and development necessary for future pest management* in *Pest management for the 21st century,* R. W. Stark and A. R. Gittens, eds. Natural Resources Series No. 2, pp. 23-38. Moscow, Idaho: Idaho Research Foundation. 102 pp.

Carlson, Gerald A., and Emery N. Castle. 1972. *Economics of pest control* in *Pest control, strategies for the future.* Div. of Biology and Agriculture. Washington, D.C.: National Academy of Sciences -National Research Council, pp. 79-99.

Chant, D. A. 1966. *Research need for integrated control.* Pages 103-10 in *Proceedings of the FAO Symposium on Integrated Pest Control, III.* (Oct. 11-15, 1965). Rome: FAO, United Nations. 129 pp.

Chiarappa, L. 1974. *Possibility of supervised plant disease control in pest management systems.* FAO Plant Protection Bull. 22: 65-68. Rome: FAO, United Nations.

Chiu, Ren-jong. 1974. Personal communication. Joint Commission on Rural Reconstruction, Taipei, Republic of China (Taiwan).

—— and David F. Yen. 1972. *Plant protection training and future needs in Taiwan.* PID-C-358 (Dec. 1, 1972). Joint Commission on Rural Reconstruction, Taipei, Taiwan.

Coon, D. W., and R. R. Fleet. 1970. *The ant war.* Environment 12(10): 28-38.

Couch, Houston B. 1973. *Current status of undergraduate programs in plant protection in the United States.* Prepared for the 8th Annual Conference, Association of Plant Pathology Dept. Chairmen, University of Minnesota (Sept. 6, 1973). (Address: Dept. of Plant Pathology and Physiology, Virginia Polytechnic Institute and State Univ., Blacksburg, Va. 24061.)

———. 1974. Personal communication. Dept. of Plant Pathology, Virginia Polytechnic Institute and State Univ., Blacksburg, Va.

Cox, R. S. 1976. Letter to the Editor. Phytopathology News 10(12): 3.

Dickerson, O. J. 1975. Personal communication. Dept. of Plant Pathology. Kansas State Univ., Manhattan, Kan.

Erickson, L. C. 1974. Personal communication. Plant Science Div., Univ. of Idaho, Moscow, Idaho.

Fitzsimmons, K. R. 1972. *Role of industry in advancing new pest control strategies.* Pages 352–61 in *Pest control, strategies for the future.* Div. of Biology and Agriculture. Washington, D.C.: National Academy of Sciences–National Research Council.

Furtick, W. R. 1975. Personal communication. Chief, Plant Protection Div., FAO, United Nations, Rome. (New address: Dean of Agriculture, Univ. of Hawaii, Honolulu, Hawaii.)

Gilbert, J. C. 1974. Personal communication. Dept. of Horticulture, Univ. of Hawaii, Honolulu, Hawaii.

Glass, Edward H. (Coordinator) 1975. *Integrated pest management: rationale, potential needs and implementation.* Entomol. Soc. of Amer. ESA Special Publication, number 75-2.

Good, J. M. 1973a. *Evolution of pest management programs.* Symposium paper: Annual Meeting, Entomol. Soc. Amer. (Nov. 26–30, 1973). Dallas, Tex.

———. 1973b. *Pilot programs for integrated pest management in the United States.* Paper presented at US–USSR Pest Management Conference (Sept. 10–18, 1973). Kiev, USSR.

———. 1974a. *Integrated pest management programs.* Paper given at Pest Management Symposium, 1974 Annual Meeting, Weed Science Society of America (Feb. 12–14, 1974). Las Vegas, Nev.

———. 1974b. *Extension emphasis and results from cotton pest management projects.* Report to 1974 Beltwide Cotton Production and Mechanization Conference (Jan. 9–10, 1974). Dallas, Tex.

———. 1974c. *Pilot pest management projects.* Statement prepared Feb. 1, 1974. Director, Pest Management Programs. Extension Service, U.S. Dept. of Agriculture, Washington, D.C.

———. 1974d. Personal communication.

———. 1975. Personal communication.

Green, Ralph J., Jr. 1974. Personal communication. Dept. of Botany and Plant Pathology, Purdue Univ., Lafayette, Ind.

Handler, Philip. 1971. *The federal government and the scientific community.* Science 171: 144–51.

Hare, Woodrow. 1974. Personal communication. Dept. of Plant Pathology, Mississippi State Univ., State College, Miss.

Harpaz, Isaac, and David Rosen. 1971. *Development of integrated control programs for crop pests in Israel.* Pages 458–63 in C. B. Huffaker, ed., *Biological Control.* Plenum Press, N.Y. 511 pp.

Heagle, Allen S. 1973. *Interactions between air pollutants and plant parasites.* Ann. Rev. Phytopathol. 11: 365–88.

Hess, Carroll. 1974. Personal communication. Dean, College of Agriculture, Kansas State Univ., Manhattan, Kan.

Hooker, A. L. 1975. Personal communication. Dept. of Plant Pathology, Univ. of Illinois, Champaign-Urbana, Ill.

Huffaker, C. B. (Project Director) 1974. *Integrated pest management. The principles, strategies and tactics of pest population regulation and control in major crop ecosystems.* Vol. 2. Detailed Institution Submittals (Sept. 1974). International Center for Biological Control. Univ. of California, Berkeley. (Report of 19 participating universities in cooperative research project.)

Janick, Jules, Robert W. Schery, Frank W. Woods, and Vernon W. Ruttan. 1969. *Plant science: an introduction to world crops.* W. H. Freeman and Co., San Francisco, Calif. 629 pp.

Jones, L. R. 1919. *Our journal, phytopathology.* Phytopathology 9: 159–64.

Kelman, Arthur. 1974. Personal communication. Dept. of Plant Pathology, Univ. of Wisconsin, Madison, Wis.

———. 1975. Personal communication.

Knutson, Herbert. 1975. Personal communication. Entomology Dept., Kansas State Univ., Manhattan, Kan.

Kung, Ku-sheng. 1975. Personal communication. Director, Plant Protection Center, Taiwan, Wufeng, Taichung, Taiwan, Republic of China.

Larson, Joseph R. 1974. Personal communication. Dept. of Entomology, Univ. of Illinois, Champaign-Urbana, Ill.

Lincoln, Charles Grover, C. Dowell, W. P. Boyer, and Robert C. Hunter. 1963. *The point sample method of scouting for boll wevil.* Ark. Agric. Expt. Sta. Bull. 666. Univ. of Arkansas, Fayetteville, Ark. 31 pp.

Lo, T. T. 1969. *Plant protection in Taiwan.* Pages 95-98 in *Thinking and doing for better crop production—PID activities.* Joint Commission on Rural Reconstruction, Plant Industry Series No. 28. Plant Industry Div. Taipei, Taiwan, Republic of China. 130 pp.

Luh, C. L., ed. 1971. *Better crop production: problems and solutions—PID activities.* Joint Commission on Rural Reconstruction, Plant Industry Series No. 30. Plant Industry Div., Taipei, Taiwan, Republic of China. 93 pp.

MacKenzie, Jake. 1975. Personal communication. Regional Chief, Pesticide Programs. U.S. Environmental Protection Agency, 100 California Street, San Francisco, Calif.

McCallan, S. E. A. 1969. *A perspective on plant pathology.* Ann. Rev. Phytopathol. 7: 1-12.

McKelvey, John J., Jr. 1975. Personal communication. The Rockefeller Foundation, N.Y.

McNew, G. L. 1972. *Concept of pest management.* Pages 119-33 in *Pest control strategies for the future.* Div. of Biology and Agriculture. Washington, D.C.: National Academy of Sciences-National Research Council.

Metcalf, C. L., W. P. Flint, and R. L. Metcalf. 1951. *Destructive and useful insects.* McGraw-Hill Book Co., N.Y. 1071 pp.

Miller, Stanley F. 1974. Personal communication. IPPC, Oregon State Univ., Corvallis, Oreg.

Monk, Richard M. 1974. Personal communication. Safety Engineer, Div. of Health, Office of Standards Development, U.S. Dept. of Labor, Washington, D.C.

Morison, R. S. 1969. *Science and social attitudes.* Science 165 (3889): 150-56.

Nusbaum, C. J., and Howard Ferris. 1973. *The role of cropping systems in nematode population management.* Ann. Rev. Phytopathol. 11: 423-40.

Osmun, John V. 1972. *Training for transition.* Pages 188-95 in *Implementing practical pest management strategies.* Proc. National Extension Insect Pest Management Workshop (Mar. 14-16, 1972). Purdue Univ., Lafayette, Ind. (Sponsored by U.S. Dept. of Agriculture Extension Service, cooperating with state extension services.)

Paddock, William C. 1967. *Phytopathology in a hungry world.* Ann. Rev. Phytopathol. 5: 375-90.

Painter, R. H. 1951. *Insect resistance in crop plants.* Macmillan, N.Y. 520 pp.

Palm, Charles E. 1972. *The role of the land grant universities in developing and implementing pest management programs.* Pages 201-206 in *Implementing practical pest management strategies.* Proc. National Extension Insect Pest Management Workshop (Mar. 14-16, 1972). Purdue Univ., Lafayette, Ind. 206 pp. (Sponsored by U.S. Dept. of Agriculture Extension Service, cooperating with state extension services.)

Pickett, A. D., and A. W. MacPhee. 1965.*Twenty years experience with integrated control programs in Nova Scotia apple orchards.* Proc. XIIth Int. Congr. Entomol., London. 597 pp.

Robins, J. S. 1972. *Role of USDA in pest management programs.* Pages 196-200 in *Implementing practical pest management strategies.* Proc. National Extension Insect Pest Management Workshop (Mar. 14-16, 1972). Purdue Univ., Lafayette, Ind. (Sponsored by U.S. Dept of Agriculture Extension Service, cooperating with the state extension services.)

Russell, Sir E. John. 1955. *The changing problems of applied biology.* Ann. Appl. Biol. 42: 8-21.

Santelman, Paul W. 1974. Personal communication. Agronomy Dept., Oklahoma State Univ., Stillwater, Okla.

Schmidt, John. 1975. Personal communication. Dept. of Agronomy, Univ. of Nebraska, Lincoln, Neb.

Schultz, Otto. 1974. *International extension advisory work.* Phytopathology News 8: 4–5.

Shaw, W. C., and L. L. Jansen. 1972. *Chemical weed control strategies for the future.* Pages 197–215 in *Pest control: strategies for the future.* Div. of Biology and Agriculture. Washington, D.C.: National Academy of Sciences–National Research Council.

Shay, J. Ralph. 1974. Personal communication. Dept. of Plant Pathology, Oregon State Univ., Corvallis, Oreg.

Sherf, Arden F. 1973. *The development and future of extension plant pathology in the United States.* Ann. Rev. Phytopathol. 11: 487–512.

Smith, Edward H. 1972. *Implementing pest control strategies.* Pages 44–68 in *Pest control: strategies for the future.* Div. of Biology and Agriculture. Washington, D.C.: National Academy of Sciences–National Research Council.

Smith, Harlan. 1973. *An analysis of insect pest management principles and how they might be applied to control plant diseases.* (Sept. 1973) Plant Pathologist, Science and Education Admin., Dept. of Agriculture, Washington, D.C. 20250.

———. 1974. Personal communication.

———.1976a. *Potentials of crop health.* Phytopathology News 10(2): 5.

———. 1976b. *List of pathology consultants available.* Personal communication.

Smith, R. F. 1969. *The new and old in pest control.* Proc. Acad. Naz. dei Lincei (Rome) (1968) 366(128): 21–30.

———. 1972. *The impact of the green revolution on plant protection in tropical and sub-tropical areas.* Bull. Entomol. Soc. of Amer. 18(1): 7–14.

———, and William W. Allen. 1954. *Insect control and the balance of nature.* Scientific American 190: 38–42.

———, and Kenneth S. Hagen. 1959. *Integrated control programs in the future of biological control.* J. Econ. Entomol. 52: 1106-1108.

Stakman, E. C. 1964. *Opportunity and obligation in plant pathology.* Ann. Rev. Phytopathol. 2: 1-12.

———, and J. George Harrar. 1957. *Principles of plant pathology.* Ronald Press, N.Y. 581 pp.

Sturgeon, R. V., Jr. 1974a. Personal communication. Dept. of Botany and Plant Pathology, Oklahoma State Univ., Stillwater, Okla.

———. 1974b. *Plant health programs.* Phytopathology News 8(3): 5-6.

Swarup, Vishnu. 1974. Personal communication. Horticulture Div. Indian Agricultural Research Institute, New Delhi, India.

Tammen, James. 1974. Personal communication. Dept. of Plant Pathology, Pennsylvania State Univ., University Park, Pa.

———. 1976. Personal communication.

Thurston, H. David. 1974. Personal communication. Dept. of Plant Pathology, Cornell Univ., Ithaca, N.Y.

Van der Plank, J. E. 1963. *Plant diseases: epidemics and control.* Academic Press, N.Y.

———. 1972. *Basic principles of ecosystems analysis.* Pages 109-118 in *Pest control: strategies for the future.* Div. of Biology and Agriculture. Washington, D.C.: National Academy of Sciences-National Research Council.

———. 1975. *Principles of plant infection.* Academic Press, London. 216 pp.

Walker, J. C. 1963. *The future of plant pathology.* Ann. Rev. Phytopathol. 1: 1-4.

Wellman, F. L. 1968. *More diseases on crops in the tropics than in the temperate zone.* Ceiba 14: 17-28.

Wharton, C. R., Jr. 1969. *The green revolution: cornucopia or Pandora's box?* Foreign Affairs 47: 464-76.

Wilcke, H. L. 1972. *The role of the food industry in solving pest control problems.* Pages 69-76 in *Pest control: strategies for the future*. Div. of Biology and Agriculture. Washington, D.C.: National Academy of Sciences-National Research Council.

Wood, David L., Robert M. Silverstein, and Minoru Nakajima. 1969. *Pest control*. Science 164(3876): 203–210.

Wood, Francis A. 1974. Personal communication. Dept. of Plant Pathology, Univ. of Minnesota, St. Paul, Minn.

Yarwood, C. E. 1968. *Tillage and plant diseases*. BioScience 18: 27–30.

———. 1970. *Man-made plant diseases*. Science 168: 218–220.

Index